# Multivariate Statistics Made Simple

## A Practical Approach

# Multivariate Statistics Made Simple

## A Practical Approach

K. V. S. Sarma

R. Vishnu Vardhan

CRC Press
Taylor & Francis Group
Boca Raton  London  New York

CRC Press is an imprint of the
Taylor & Francis Group, an **informa** business

CRC Press
Taylor & Francis Group
6000 Broken Sound Parkway NW, Suite 300
Boca Raton, FL 33487-2742

Printed on acid-free paper
Version Date: 20180913

International Standard Book Number-13: 978-1-1386-1095-8 (Hardback)

**Visit the Taylor & Francis Web site at**
**http://www.taylorandfrancis.com**

**and the CRC Press Web site at**
**http://www.crcpress.com**

*To my beloved wife,*
*Late Sarada Kumari (1963–2015),*
*who inspired me to inspire data scientists.*
*K.V.S.Sarma*

*To my beloved parents,*
*Smt. R. Sunanda & Sri. R. Raghavendra Rao,*
*for having unshakeable confidence in my endeavors.*
*R.Vishnu Vardhan*

# Contents

Preface           xi

Authors          xv

**1   Multivariate Statistical Analysis—An Overview      1**

    1.1    An Appraisal of Statistical Analysis . . . . . . . . . . . . .   1
    1.2    Structure of Multivariate Problems   . . . . . . . . . . . .   4
    1.3    Univariate Data Description   . . . . . . . . . . . . . . . .   7
    1.4    Standard Error (SE) and Confidence Interval (CI) . . . . . .   7
    1.5    Multivariate Descriptive Statistics . . . . . . . . . . . . . .   8
    1.6    Covariance Matrix and Correlation Matrix   . . . . . . . . .   14
    1.7    Data Visualization . . . . . . . . . . . . . . . . . . . . . .   16
    1.8    The Multivariate Normal Distribution . . . . . . . . . . . .   23
    1.9    Some Interesting Applications of Multivariate Analysis   . . .   25
    Summary . . . . . . . . . . . . . . . . . . . . . . . . . . . .   26
    Do it yourself   . . . . . . . . . . . . . . . . . . . . . . . .   27
    Suggested Reading . . . . . . . . . . . . . . . . . . . . . . .   29

**2   Comparison of Multivariate Means      31**

    2.1    Multivariate Comparison of Mean Vectors . . . . . . . . . .   31
    2.2    One-sample Hotelling's $T^2$ Test   . . . . . . . . . . . . . .   33
    2.3    Confidence Intervals for Component Means . . . . . . . . . .   37
    2.4    Two-Sample Hotelling's $T^2$ Test   . . . . . . . . . . . . . .   39
    2.5    Paired Comparison of Multivariate Mean Vectors   . . . . .   43
    Summary . . . . . . . . . . . . . . . . . . . . . . . . . . . .   48
    Do it yourself   . . . . . . . . . . . . . . . . . . . . . . . .   49
    Suggested Reading . . . . . . . . . . . . . . . . . . . . . . .   50

**3   Analysis of Variance with Multiple Factors      51**

    3.1    Review of Univariate Analysis of Variance (ANOVA)   . . . .   51
    3.2    Multifactor ANOVA . . . . . . . . . . . . . . . . . . . . .   55
    3.3    ANOVA with a General Linear Model . . . . . . . . . . . .   57
    3.4    Continuous Covariates and Adjustment   . . . . . . . . . . .   61

3.5   Non-Parametric Approach to ANOVA . . . . . . . . . . . .   65
3.6   Influence of Random Effects on ANOVA   . . . . . . . . . .   68
Summary . . . . . . . . . . . . . . . . . . . . . . . . . . . . .   69
Do it yourself   . . . . . . . . . . . . . . . . . . . . . . . . .   69
Suggested Reading . . . . . . . . . . . . . . . . . . . . . . . .   71

**4   Multivariate Analysis of Variance (MANOVA)            73**

4.1   Simultaneous ANOVA of Several Outcome Variables . . . . .   73
4.2   Test Procedure for MANOVA   . . . . . . . . . . . . . . . .   74
4.3   Interpreting the Output . . . . . . . . . . . . . . . . . . .   78
4.4   MANOVA with Age as a Covariate   . . . . . . . . . . . . .   81
4.5   Using Age and BMI as Covariates   . . . . . . . . . . . . .   84
4.6   Theoretical Model for Prediction . . . . . . . . . . . . . .   86
Summary . . . . . . . . . . . . . . . . . . . . . . . . . . . . .   88
Do it yourself   . . . . . . . . . . . . . . . . . . . . . . . . .   88
Suggested Reading . . . . . . . . . . . . . . . . . . . . . . . .   90

**5   Analysis of Repeated Measures Data                    91**

5.1   Experiments with Repeated Measures . . . . . . . . . . . .   91
5.2   RM ANOVA Using SPSS . . . . . . . . . . . . . . . . . . .   95
5.3   RM ANOVA Using MedCalc . . . . . . . . . . . . . . . . .   99
5.4   RM ANOVA with One Grouping Factor . . . . . . . . . . .   100
5.5   Profile Analysis   . . . . . . . . . . . . . . . . . . . . . .   107
Summary . . . . . . . . . . . . . . . . . . . . . . . . . . . . .   113
Do it yourself   . . . . . . . . . . . . . . . . . . . . . . . . .   113
Suggested Reading . . . . . . . . . . . . . . . . . . . . . . . .   114

**6   Multiple Linear Regression Analysis                   115**

6.1   The Concept of Regression   . . . . . . . . . . . . . . . . .   115
6.2   Multiple Linear Regression   . . . . . . . . . . . . . . . . .   118
6.3   Selection of Appropriate Variables into the Model   . . . . .   120
6.4   Predicted Values from the Model . . . . . . . . . . . . . . .   126
6.5   Quality of Residuals . . . . . . . . . . . . . . . . . . . . .   128
6.6   Regression Model with Selected Records   . . . . . . . . . .   129
Summary . . . . . . . . . . . . . . . . . . . . . . . . . . . . .   131
Do it yourself   . . . . . . . . . . . . . . . . . . . . . . . . .   131
Suggested Reading . . . . . . . . . . . . . . . . . . . . . . . .   134

**7   Classification Problems in Medical Diagnosis          135**

7.1   The Nature of Classification Problems . . . . . . . . . . . .   135
7.2   Binary Classifiers and Evaluation of Outcomes   . . . . . . .   137
7.3   Performance Measures of Classifiers   . . . . . . . . . . . .   138

7.4 ROC Curve Analysis . . . . . . . . . . . . . . . . . . . . . 144
7.5 Composite Classifiers . . . . . . . . . . . . . . . . . . . . 148
7.6 Biomarker Panels and Longitudinal Markers . . . . . . . 149
Summary . . . . . . . . . . . . . . . . . . . . . . . . . . . . . 150
Do it yourself . . . . . . . . . . . . . . . . . . . . . . . . . . 150
Suggested Reading . . . . . . . . . . . . . . . . . . . . . . . . 151

**8 Binary Classification with Linear Discriminant Analysis    153**

8.1 The Problem of Discrimination . . . . . . . . . . . . . . . 153
8.2 The Discriminant Score and Decision Rule . . . . . . . . . 155
8.3 Understanding the Output . . . . . . . . . . . . . . . . . . 158
8.4 ROC Curve Analysis of Discriminant Score . . . . . . . . 161
8.5 Extension of Binary Classification . . . . . . . . . . . . . 163
Summary . . . . . . . . . . . . . . . . . . . . . . . . . . . . . 164
Do it yourself . . . . . . . . . . . . . . . . . . . . . . . . . . 164
Suggested Reading . . . . . . . . . . . . . . . . . . . . . . . . 167

**9 Logistic Regression for Binary Classification    169**

9.1 Introduction . . . . . . . . . . . . . . . . . . . . . . . . . 169
9.2 Simple Binary Logistic Regression . . . . . . . . . . . . . 170
9.3 Binary Logistic Regression with Multiple Predictors . . . . 175
9.4 Assessment of the Model and Relative Effectiveness of
     Markers . . . . . . . . . . . . . . . . . . . . . . . . . . . . 179
9.5 Logistic Regression with Interaction Terms . . . . . . . . . 180
Summary . . . . . . . . . . . . . . . . . . . . . . . . . . . . . 182
Do it yourself . . . . . . . . . . . . . . . . . . . . . . . . . . 182
Suggested Reading . . . . . . . . . . . . . . . . . . . . . . . . 184

**10 Survival Analysis and Cox Regression    185**

10.1 Introduction . . . . . . . . . . . . . . . . . . . . . . . . . 185
10.2 Data Requirements for Survival Analysis . . . . . . . . . . 186
10.3 Estimation of Survival Time with Complete Data
      (No Censoring) . . . . . . . . . . . . . . . . . . . . . . . . 187
10.4 The Kaplan–Meier Method for Censored Data . . . . . . . 190
10.5 Cox Regression Model . . . . . . . . . . . . . . . . . . . . 193
Summary . . . . . . . . . . . . . . . . . . . . . . . . . . . . . 199
Do it yourself . . . . . . . . . . . . . . . . . . . . . . . . . . 200
Suggested Reading . . . . . . . . . . . . . . . . . . . . . . . . 203

## 11 Poisson Regression Analysis                                      205

11.1 Introduction . . . . . . . . . . . . . . . . . . . . . . . . . 205
11.2 General Form of Poisson Regression . . . . . . . . . . . . . 206
11.3 Selection of Variables and Subset Regression . . . . . . . . 211
11.4 Poisson Regression Using SPSS . . . . . . . . . . . . . . . 213
11.5 Applications of the Poisson Regression model . . . . . . . . 216
Summary . . . . . . . . . . . . . . . . . . . . . . . . . . . . . 217
Do it yourself . . . . . . . . . . . . . . . . . . . . . . . . . . 217
Suggested Reading . . . . . . . . . . . . . . . . . . . . . . . . 218

## 12 Cluster Analysis and Its Applications                            219

12.1 Data Complexity and the Need for Clustering . . . . . . . . 219
12.2 Measures of Similarity and General Approach to Clustering . 221
12.3 Hierarchical Clustering and Dendrograms . . . . . . . . . . 224
12.4 The Impact of Clustering Methods on the Results . . . . . . 229
12.5 K-Means Clustering . . . . . . . . . . . . . . . . . . . . . 232
Summary . . . . . . . . . . . . . . . . . . . . . . . . . . . . . 235
Do it yourself . . . . . . . . . . . . . . . . . . . . . . . . . . 235
Suggested Reading . . . . . . . . . . . . . . . . . . . . . . . . 237

## Index                                                                239

# Preface

The twenty-first century is recognized as the era of data science. Evidence-based decision making is the order of the day and the common man is continuously educated about *various types of information.* Statistics is a science of data analysis and a powerful tool in decision making. The days are gone when only numbers were considered as data but with the advent of sophisticated software, data now includes text, voice,and images in addition to numbers. Practical data will be usually unstructured and sometimes contaminated and hence statistics alone can resolve issues of uncertainty.

Students taking a master's course in statistics, by and large learn the mathematical treatment of statistical methods in addition to a few applications areas like biology, public health, psychology, marketing etc. However, real-world problems are often multivariate in nature and the tools needed are not as simple as describing a data using averages, measures of spread or charts.

Multivariate statistical analysis requires the ability to handle complex mathematical routines such as inverse of matrices, eigen values or complicated integrals of density functions and lengthy iterations. Statistical software has transformed analytic tools into simple and user-friendly applications, thereby bridging the gap between science (theory) and the technology of statistics. Expertise in statistical software is inevitable to the contemporary statisticians. Easy access to software packages however distracts the attention of users away from 'a state of understanding' to a 'cookbook mode' focusing on the end results only. The current tendency towards reporting p-values is one such situation without rational justification for the context of use.

On the other hand, there is a tremendous growth in data warehousing, data mining and machine learning. Data analytic tools like artificial intelligence, image processing, dynamic visualization of data etc., have shadowed the classical approach to statistical analysis and real-time outputs are available on the dashboards of computer-supported devices.

The motivation for publishing this book was our experience in teaching statistics to post-graduate students over a long period of time as well as interaction with professionals including medical researchers, practicing doctors, psychologists, management experts and engineers. The problems brought to us are often challenging and need a different view when compared with what is taught in a classroom. The selection of suitable statistical tools (like Analysis of Variance) and an implementation tool (like software or an online calculator) plays an important role for the application scientist. Hence it is time to learn and practice statistics by utilizing the power of computers. Our experience in attending to problems in consulting also made us to write this book.

Teaching and learning statistics is fruitful when real-time studies are understood. We have chosen medicine and clinical research areas as the platform to explain statistics tools. The discussions and illustrations however hold well for other contexts too.

All the illustrations are explained with one or more of the following software packages.

1. IBM SPSS Statistics Version 20 (IBM Corp, Somers, NY, USA)

2. MedCalc Version 15.0 for Windows (MedCalc Software bvba, Belgium)

3. Real Statistics Resource Pack Add-ins to Microsoft Excel

4. R (open source software)

This book begins with an overview of multivariate analysis in Chapter 1 designed to motivate the reader towards a 'holistic approach' for analysis instead of partial study using univariate analysis. In Chapter 2 we have focused on the need to understand several related variables simultaneously, using mean vectors. The concept of Hotelling's T-square and the method of working on it with MS-Excel Add-ins is illustrated. In Chapter 3 we have illustrated the use of a General Linear Model to handle Multifactor ANOVA with interactions and continuous covariates. As an extension to ANOVA we have discussed the method and provided working skills to perform Multivariate ANOVA (MANOVA) in Chapter 4 and illustrated it with practical data sets. Repeated Measures data is commonly used in follow-up studies and Chapter 5 is designed to explain the theory and skills of this tool. In Chapter 6 Multiple Linear Regression and its manifestations are discussed with practical data sets and explained with appropriate software.

An important area in medical research is the design of biomarkers for classification and early detection of diseases. Chapter 7 is devoted to explaining the statistical methods and software support for medical diagnosis and ROC curves. Linear Discriminant Analysis, a popular tool for binary classification is discussed in Chapter 8 and the procedure is illustrated with clinical data. The use of Logistic Regression in binary classification is common in predictive models and the same is discussed in Chapter 9 with practical datasets. In Chapter 10 the basic elements of Survival Analysis, the use of Kaplan–Meier test and Cox Regression are presented in view of their wide applications in the management of chronic diseases. Poisson Regression, a tool so far discussed in a few advanced books only, is presented in Chapter 11 with a focus on application and software supported skills. We end the book with Cluster Analysis in Chapter 12 in view of its vast applications in community medicine, public health and hospital management.

At the end of each chapter, we have provided exercises under the caption, Do it Yourself. We believe that the user can perform the required analysis by following the stepwise methods and the software options provided in the text. These problems are not simply numerically focused but contain a real context. A few motivational and thought-provoking exercises are also given.

We acknowledge the excellent support extended by several doctors and clinical researchers from Sri Venkateswara Institute of Medical Sciences (SVIMS), Tirupati and those at Jawaharlal Institute of Post Graduate Medical Education & Research (JIPMER), Pondicherry and the faculty members of Sri Venkateswara University, Tirupati and Pondicherry University, Puducherry. We appreciate their intent in sharing their datasets to support the illustration, in making the book more practical than a mere theoretical volume. We have acknowledged each of them in the appropriate places.

Our special thanks to Dr. Alladi Mohan, Professor and Head, Department of Medicine, Sri Venkateswara Institute of Medical Sciences (SVIMS), Tirupati for lively discussion on various concepts used in this book.

We also wish to place on record our special appreciation to Sai Sarada Vedururu, Jahnavi Merupula and Mohammed Hisham who assisted in database handling, running of specific tools and the major activity of putting the text into Latex format.

K.V.S.Sarma
R.Vishnu Vardhan

# Authors

### Dr. K. V. S. Sarma

K.V.S.Sarma served as a Professor of Statistics at Sri
Venkateswara University from 1977 till 2016. He taught
various subjects in statistics for 39 years including Operations Research, Statistical Quality Control, Probability
Theory, and Computer Programming. His areas of research interests are Inventory Modeling, Sampling Plans
and Data Analysis. He has guided 20 Ph.D. theses on
diverse topics and published more than 60 research articles in reputed international journals. He has authored
a book titled *Statistics Made Simple* using Excel and
SPSS, apart from delivering lectures on software-based
analysis and interpretation. He has delivered more than
100 invited talks at various conferences and training programs and he is a life
member of the Indian Society for Probability and Statistics (ISPS) and the
Operations Research Society of India. He is a consultant to clinical researchers
and currently working as Biostatistician cum Research Coordinator at the Sri
Venkateswara Institute of Medical Sciences (SVIMS).

### Dr. R. Vishnu Vardhan

Rudravaram Vishnu Vardhan is currently working as
Assistant Professor in the Department of Statistics,
Pondicherry Central University, Puducherry. His areas
of research are Biostatistics - Classification Techniques;
Multivariate Analysis; Regression Diagnostics and Statistical Computing. He has published 52 research papers
in reputed national and international journals. He has
guided two Ph.D. students and 39 P.G. projects. He is
a recipient of the Ms. Bhargavi Rao and Padma Vib-
hushan Prof. C. R. Rao Award for best Poster Presenta-
tion in an International Conference in the year 2010, Indian Society for Probability and Statistics (ISPS) *Young
Statistician Award*, December 2011 and *Young Scientist Award* from the Indian Science Congress, February 2014. He has authored a book and also edited

a book. He has presented 34 research papers in national and international conferences/Seminars, and delivered 39 invited talks at various reputed institutes/universities in India. He has organized three national workshops, one national conference and one international conference. He is a life member of several professional organizations. He serves as referee and also holds the position of editorial member at reputed journals.

# Chapter 1

# Multivariate Statistical Analysis—An Overview

| | | |
|---|---|---|
| 1.1 | An Appraisal of Statistical Analysis .............................. | 1 |
| 1.2 | Structure of Multivariate Problems ............................. | 4 |
| 1.3 | Univariate Data Description ..................................... | 7 |
| 1.4 | Standard Error (SE) and Confidence Interval (CI) ............. | 7 |
| 1.5 | Multivariate Descriptive Statistics ............................. | 8 |
| 1.6 | Covariance Matrix and Correlation Matrix ..................... | 14 |
| 1.7 | Data Visualization ............................................. | 16 |
| 1.8 | The Multivariate Normal Distribution ......................... | 23 |
| 1.9 | Some Interesting Applications of Multivariate Analysis ........ | 25 |
| | Summary ...................................................... | 26 |
| | Do it yourself (Exercises) ..................................... | 27 |
| | Suggested Reading ............................................. | 29 |

I only believe in statistics that I doctored myself.

Galileo Galilei (1564 – 1642)

## 1.1   An Appraisal of Statistical Analysis

Statistical analysis is an essential component in the contemporary business environment as well as in scientific research. With expanding horizons in all fields of research, new challenges are faced by researchers. The classical statistical analysis which is mainly focused on drawing inferences on the parameters of a population requires a conceptual change towards an exploratory study of large and complex data sets, analyzing them and mining the latent features of the data.

Statistical analysis is generally carried out with two purposes:

1. To describe the observed data for an understanding of the facts and

2. To draw inferences for the target group, (called *population*) based on sample data

The former is known as *descriptive statistics* and latter is called *inferential statistics*.

Since it is impossible or sometimes prohibitive to study the entire population, one prefers to take a *sample* and measure some characteristics or observe attributes like color, taste, preference etc. The sample data shall however be unbiased and represent the population so that the inferences drawn from the sample can be generalized to the population usually with some error. The alternative to avoid error will be examining the entire population (known as *census*), which is not always meaningful, like a proposal to pump out the entire blood from a patient's body to calculate the glucose level!

Large volumes of data with hundreds of variables and several thousands of records constitute today's data sets and the solutions are required in real time. Studying the relationships among the variables provides useful information for decision making with the help of multivariate analytical tools which refers to the "simultaneous analysis of several inter-related variables".

In a different scenario, a biologist may wish to estimate the toxic effect of a treatment on different organs of an animal. With the help of statistically designed experiments, one can estimate the effect of various prognostic factors on the response variables simultaneously by using multivariate analysis.

In general the multivariate approach is helpful to:

- Explore the joint performance of several variables and

- Estimate the effect of each variable in the presence of the others (marginal effect).

The science of multivariate of analysis is mathematically complicated and the computations are too involved to perform with a desktop calculator. However, with the availability of computer software, the burden of computation is minimized, irrespective of the size of the data. For this reason, during the last three decades multivariate tools have reached the application scientist and multivariate analysis is the order of the day.

Data analysts and some researchers use the terms *univariate* and *multivariate analyses* with the second one usually followed by the first. A brief description is given below of these approaches.

## Univariate analysis:

Analysis of variables one at a time (each one separately) is known as *univariate analysis* in which data is described in terms of mean, mode, median, standard deviation and also by way of charts. Inferences are also drawn separately for each variable, such as comparison of mean values of each variable across two or more groups of cases.

Here are some instances where univariate analysis alone is carried out.

- Daily number of MRI scans handled at a hospital

- Arrival pattern of cases at the emergency department of a hospital

- Residual life (in months) after a cancer therapy in a follow-up study

- Waiting time at toll gates

- Analyzing pain scores of patients after some intervention like Visual Analogue Scale (VAS)

- Symptomatic assessment of the health condition of a patient such as presence of fever (yes or no) or diabetic status

Each characteristic is denoted by a random variable X having a known statistical distribution. A distribution is a mathematical formula used to express the pattern of occurrence of values of X in the presence of an uncertain environment, guided by a probability mechanism.

Normal distribution is one such probability distribution used to describe a data pattern of values measured on an interval scale. The normal distribution is described by two parameters, viz., mean ($\mu$) and variance ($\sigma^2$). When the values of the parameters are not known in advance (from prior studies or by belief) they are estimated from sample data. One or more *hypotheses* can also be *tested* for their statistical significance based on the sample data. The word *statistical significance* is used to mean that the *finding from the study* is not an occurrence by chance. This area of analysis is called *inferential statistics*.

## Multivariate analysis:

Analysis of data simultaneously on several variables (characteristics) is often called *multivariate analysis*. In clinical examination, a battery of parameters (like lipid profile or hemogram) is observed on a single patient to understand a health condition. When only two variables are involved, it is called *bivariate analysis* in which, correlations, associations and simple relationships (like dose-response) are studied. Multivariate analysis is however more general and covers bivariate and univariate analyses within.

Here are some instances that fit into a multivariate environment.

- Anthropometry of a patient (height, weight, waist circumference, waist-hip ratio etc.)

- Scores obtained on Knowledge, Adoption and Practice (KAP) measured on agricultural farmers

- Response to several questions regarding pain relief after physiotherapy (having say 50 questions with response on a 5 – point scale)

- Health profile of post menopausal women (a mixture of questions on different scales)

- Repeated measurements like blood sugar levels taken for the same patient at different time points

Understanding several characteristics, in relation to others, is the essence of a multivariate approach. A group of characteristics on an individual taken together convey more information about the features of the individual than each characteristic separately. Hence several variables will be considered as a group or an *array* and such arrays are compared across study groups in multivariate analysis.

Further, data on several related variables are combined in a *suitable* way to produce a new value like a *composite score* to estimate an outcome. This however needs extra mathematical effort to handle complex data structures and needs a different approach than the univariate method.

In the following section, a brief sketch of multivariate problems is discussed.

---

## 1.2   Structure of Multivariate Problems

Suppose there are k-variables on which data is observed from an individual and let there be n-individuals in the study. Each variable is in fact a random variable to mean that the true value on the individual is unknown but governed by a random mechanism that can be explained by rules of probability.

The data can be arranged as an *array* or a *profile* denoted by $\mathbf{X} = \begin{pmatrix} X_1 \\ X_2 \\ \vdots \\ X_k \end{pmatrix}$

For instance scores on X = {Knowledge, Adoption, Practice} is a profile. The *complete hemogram* of a patient is another example of a profile. The data obtained from n individuals on X contains information on each of the p components and can be arranged as a (n x k) array or matrix shown as

$$A = \begin{pmatrix} X_{11} & X_{12} & \cdots & X_{1k} \\ X_{21} & X_{22} & \cdots & X_{2k} \\ \cdots & \cdots & \cdots & \cdots \\ X_{n1} & X_{n2} & \cdots & X_{nk} \end{pmatrix}$$

Lowercase letters are used to indicate the values obtained from the corresponding variable $X_{ij}$. For instance $x_{23}$ indicates the value on the variable $X_3$ from the individual 2.

Consider the following illustration.

**Illustration 1.1** A clinical researcher has collected data on several characteristics listed below from 64 patients with a view to compare the parameters between cases and controls. Thirty-two patients with Rheumatoid Arthritis (RA) are classified as 'cases' and thirty-two age and gender matched healthy subjects are classified as 'controls'. This categorization variable is shown as 'Group'. The variables used in the study and their description is given below.

| Variable | Parameter | Description | Type |
|---|---|---|---|
| X1 | Group | Case/Control(1/0) | Nominal |
| X2 | AS | Atherosclerosis (AS: 1 = Yes, 0 = No) | Nominal |
| X3 | Age | Age of the patient in years | Scale |
| X4 | Gen | Gender (0 = Male, 1 = Female) | Nominal |
| X5 | BMI | Body Mass Index ($Kg/m^2$) | Nominal |
| X6 | CHOL | Cholesterol (mg/dl) | Scale |
| X7 | TRIG | Triglycerides (mg/dl) | Scale |
| X8 | HDL | High-density lipoproteins cholesterol (mg/dl) | Scale |
| X9 | LDL | Low-density lipoprotein cholesterol (mg/dl) | Scale |
| X10 | VLDL | Very-low-density lipoprotein cholesterol (mg/dl) | Scale |
| X11 | CIMT | Carotid Intima-Media Thickness | Scale |

Table 1.1 shows a sample of 20 records from the study. The analysis and discussion is however based on the complete data. This data will be referred to as 'CIMT data' for further reference.

There are 11 variables out of which some (like CHOL and BMI) are measured on an interval scale while some (like Sex and Diagnosis) are coded as 0 and 1 on a nominal scale. The measured variables are called *continuous variables* and data on such variables is often called *quantitative data* by some researchers.

TABLE 1.1: Illustrative multivariate data on CIMT

| S.No | X1 | X2 | X3 | X4 | X5 | X6 | X7 | X8 | X9 | X10 | X11 |
|------|----|----|----|----|-------|-----|-----|----|-------|------|-------|
| 1 | 1 | 1 | 56 | 1 | 29.55 | 160 | 91 | 29 | 112.8 | 18.2 | 0.600 |
| 2 | 1 | 1 | 43 | 0 | 17.33 | 190 | 176 | 31 | 123.8 | 35.2 | 0.570 |
| 3 | 1 | 1 | 45 | 0 | 29.08 | 147 | 182 | 32 | 76.6 | 36.4 | 0.615 |
| 4 | 1 | 1 | 39 | 0 | 24.03 | 168 | 121 | 38 | 105.6 | 24.4 | 0.630 |
| 5 | 1 | 0 | 31 | 0 | 25.63 | 146 | 212 | 38 | 65.6 | 42.4 | 0.470 |
| 6 | 1 | 1 | 34 | 0 | 20.54 | 160 | 138 | 36 | 96.4 | 27.6 | 0.630 |
| 7 | 1 | 0 | 37 | 1 | 18.42 | 128 | 156 | 32 | 64.8 | 31.2 | 0.515 |
| 8 | 1 | 1 | 54 | 0 | 25.53 | 212 | 307 | 36 | 114.6 | 61.4 | 0.585 |
| 9 | 1 | 1 | 59 | 0 | 22.67 | 149 | 89 | 38 | 93.2 | 17.8 | 0.615 |
| 10 | 1 | 0 | 33 | 0 | 21.92 | 147 | 95 | 32 | 96.0 | 19.0 | 0.415 |
| 11 | 0 | 0 | 45 | 0 | 23.44 | 122 | 76 | 32 | 74.8 | 15.2 | 0.460 |
| 12 | 0 | 0 | 37 | 0 | 27.48 | 124 | 84 | 42 | 65.2 | 16.8 | 0.440 |
| 13 | 0 | 0 | 63 | 0 | 19.55 | 267 | 153 | 45 | 190.4 | 30.6 | 0.530 |
| 14 | 0 | 0 | 41 | 0 | 23.43 | 175 | 261 | 38 | 84.8 | 52.2 | 0.510 |
| 15 | 0 | 0 | 51 | 0 | 24.44 | 198 | 193 | 35 | 124.4 | 38.6 | 0.410 |
| 16 | 0 | 0 | 40 | 0 | 26.56 | 160 | 79 | 36 | 108.2 | 15.8 | 0.560 |
| 17 | 0 | 0 | 42 | 0 | 25.53 | 130 | 95 | 32 | 79.0 | 19.0 | 0.480 |
| 18 | 0 | 0 | 45 | 1 | 26.44 | 188 | 261 | 32 | 103.6 | 52.4 | 0.510 |
| 19 | 0 | 0 | 38 | 0 | 25.78 | 120 | 79 | 32 | 71.2 | 15.8 | 0.510 |
| 20 | 0 | 1 | 43 | 1 | 19.70 | 204 | 79 | 42 | 146.2 | 15.8 | 0.680 |

*(Data courtesy: Dr. Alladi Mohan, Department of Medicine, Sri Venkateswara Institute of Medical Sciences (SVIMS), Tirupati.)*

The other variables which are not measured on a scale are said to be *discrete* and the word *qualitative data* is used to indicate them. It is important to note that means and standard deviations are calculated only for quantitative data while counts and percentages are used for qualitative data.

The rows represent the *records* and the columns represent the *variables*. While the records are independent, the variables need not be. In fact, the data on several variables (columns) exhibit related information and multivariate analysis helps summarize the data and draw inferences about *sets of parameters* (of variables) instead of a single parameter at a time.

Often, in public health studies there will be a large number of variables used to describe one context. During analysis, some new variables will also be created like a partial sum of scores, derived values like BMI and so on.

The size *(n × k)* of the data array is called the dimensionality of the problem and the complexity of analysis usually increases with *k*. This type of issue often arises in health studies and also in marketing.

In the following section we outline the tools used for univariate description of variables in terms of average, spread etc.

---

## 1.3   Univariate Data Description

For continuous variables like BMI, Weight etc., the simple average (mean) is a good summary provided that there are no extreme values in the data. It is easy to see that mean is quickly perturbed by very high or very low values in the data. Such high or low values may sometimes be important in the study their presence along with other values not only gives an incorrect average however, but also inflates the standard deviation (S.D). Some computer packages use Std. Dev. or S.D.

For instance, the data set 3,1,2,1,4,3 has a mean of 2.33 and a standard deviation of 1.21. Suppose the first and last values are modified and the new data set is -2,1,2,1,4,8. This data also has the same average of 2.33 but a standard deviation of 3.38 which shows that the second data set has more spread around the mean than the first one.

Normally distributed data will be summarized as mean $\pm$ S.D. If $Q_3$ and $Q_1$ represent the third and first quartiles (75% and 25% percentiles) respectively, then IQR $= (Q_3 - Q_1)$ is the measure of dispersion used in place of S.D. For non-symmetric data distribution (age or residual life after treatment), data will be summarized as median and IQR.

For qualitative data like gender, diabetes (yes/no), hypertension (yes/no) etc., the values are discrete and hence the concept of mean or median is not used to summarize the data. Counts and percentages are used to describe such data expressed on an ordinal or nominal scale instead of an interval scale.

In the next section we discuss the concept and utility of the confidence interval in drawing inferences about the population.

---

## 1.4   Standard Error (SE) and Confidence Interval (CI)

Suppose $\bar{x}$ denotes the sample mean of a group of n individuals on a parameter and let 's' be the S.D of the sample. There are several ways of obtaining samples of size 'n' from a given cohort or target group. So every sample gives a 'mean' which is an estimate of the true but unknown mean ($\mu$) of the cohort.

Let $\overline{x}_1, \overline{x}_2, \cdots, \overline{x}_m$ be the means of m-samples. Then the distribution (pattern) of $\overline{x}_1, \overline{x}_2, \cdots, \overline{x}_m$ is called the *sampling distribution of the mean* and the S.D of these distribution is called *Standard Error* denoted by SE. Some software report this as Std. Error.

For normally distributed data the mean (average) of $\overline{x}_1, \overline{x}_2, \cdots, \overline{x}_m$ is an estimate of $\mu$. We say that $\overline{x}$ is an *unbiased estimator* of $\mu$ and it is called a *point estimate*. It can be shown that the SE of mean is estimated as $\dfrac{s}{\sqrt{n}}$.

It is customary to provide an interval (rather than a simple value) around $\overline{x}$ such that the true value of $\mu$ lies in this interval with $100(1-\alpha)\%$ confidence where $\alpha$ is called the *error rate*. Such an interval is called the *Confidence Interval* (CI) given by

$$\left[ \overline{x} \pm Z_\alpha \frac{s}{\sqrt{n}} \right]$$

In general the CI takes the form [sample value $\pm Z_\alpha * \text{SE}$]. As a convention we take $\alpha = 0.05$ so that the 95% CI for mean will be

$$\left[ \overline{x} \pm 1.96 \frac{s}{\sqrt{n}} \right]$$

because $Z = 1.96$ for normal distribution (use NORMSINV(0.025) in MS-Excel).

CI plays a pivotal role in statistical inference. It is important to note that every estimate (sample value) contains some error, expressed in terms of SE and CI.

The following section contains methods of describing multivariate data.

---

## 1.5   Multivariate Descriptive Statistics

When the multivariate data is viewed as a collection of variables, each of them can be summarized in terms of descriptive statistics known as *mean vector* and *variance-covariance matrix*. For the data array X we define the following.

**Mean vector:**

This is defined as $\overline{x} = \begin{pmatrix} \overline{x}_1 \\ \overline{x}_2 \\ \vdots \\ \overline{x}_k \end{pmatrix}$

where $\bar{x}_j = \dfrac{1}{n} \sum_{i=1}^{n} x_{ij}$ for $j = 1, 2, \ldots, k$ $\qquad$ (1.1)

(mean of all values for the $j^{th}$ column) is the sample mean of $X_j$. Mean is expressed in the original measurement units like centimeters, grams etc.

**Variance-Covariance matrix:**

Each variable $X_j$ will have some variance that measures the spread of values around its mean. This is given by the *sample variance*.

$$s_j^2 = \frac{1}{(n-1)} \sum_{i=1}^{n} (x_{ij} - \bar{x}_j)^2 \text{ for } j = 1, 2, \ldots, k \qquad (1.2)$$

The denominator in Equation 1.2 should have been n but the use of $(n-1)$ is to correct for the small sample bias (specifically for normally distributed data) and gives what is called an *unbiased* estimate of the population mean for this variable. Most of the computer codes use Equation 1.2 since it works for both small and large n. Variance will have squared units like $cm^2$, $gm^2$ etc., which is difficult to interpret along with mean. Hence, another measure called the *standard deviation* is used which is expressed in natural units and always non-negative. When all the data values are the same, then the standard deviation is zero. The *sample standard deviation* of $X_j$ is given by

$$s_j = \sqrt{\frac{1}{(n-1)} \sum_{i=1}^{n} (x_{ij} - \bar{x}_j)^2} \text{ for } j = 1, 2, \ldots, k \qquad (1.3)$$

Another measure used in multivariate analysis is the measure of co-variation between each pair of variables, called the *covariance*. If $X_j$ and $X_k$ constitute a pair of variables then the covariance is defined as

$$s_{jk} = \frac{1}{(n-1)} \sum_{i=1}^{n} (x_{ij} - \bar{x}_j)(x_{il} - \bar{x}_l) \text{ for } j = 1, 2, \ldots, k \; l = 1, 2, \ldots, k \quad (1.4)$$

The covariance is also called the *product moment* and measures the simultaneous variation in $X_j$ and $X_l$. There will be $\dfrac{k(k-1)}{2}$ covariances for the vector X. From Equation 1.4 it is easy to see that $s_{jl}$ is the same as $s_{lj}$. The covariance between $X_j$ and $X_l$ is nothing but the variance of $X_j$.

The variances and covariances can be arranged in the form of a matrix called the *variance-covariance* matrix as follows.

$$S = \begin{pmatrix} s_1^2 & s_{12} & \cdots & s_{1k} \\ s_{21} & s_2^2 & \cdots & s_{2k} \\ \cdots & \cdots & \cdots & \cdots \\ s_{k1} & s_{k2} & \cdots & s_k^2 \end{pmatrix}$$

We also use the notation $s_j^2 = s_{jj}$ so that every element is viewed as a covariance and we write

$$S = \begin{pmatrix} s_{11} & s_{12} & \cdots & s_{1k} \\ s_{21} & s_{22} & \cdots & s_{2k} \\ \cdots & \cdots & \cdots & \cdots \\ s_{k1} & s_{k2} & \cdots & s_{nk} \end{pmatrix} \tag{1.5}$$

When each variable is studied separately, it will have only variance and the concept of covariance does not arise. The overall spread of data around the mean vector can be expressed as given below by a single metric which is useful for comparison of multivariate vectors.

1. *The generalized sample variance* defined as the determinant of the covariance matrix ($|S|$)

2. *Total sample variance* defined as the sum of all diagonal terms in the matrix $S$. This is also called *trace*, given by $\mathrm{tr}(S) = s_{11} + s_{22} + \ldots + s_{kk}$

**Correlation matrix:**

The important descriptive statistic in multivariate analysis is the sample *correlation coefficient*, denoted by $r_{jk}$ between $X_j$ and $X_k$ known as the *Pearson's product-moment correlation coefficient* proposed by Karl Pearson (1867–1936). It is a measure of the strength of the linear relationship between a pair of variables. It is calculated as

$$r_{jl} = \frac{s_{jl}}{\sqrt{s_{jj}s_{ll}}} = \frac{\sum\limits_{i=1}^{n}(x_{ij} - \bar{x}_j)(x_{il} - \bar{x}_l)}{\sqrt{\sum\limits_{i=1}^{n}(x_{ij} - \bar{x}_i^2)(\sum\limits_{i=1}^{n}(x_{il} - \bar{x}_l^2)}} \tag{1.6}$$

We sometimes write this simply as $r$ without any subscript. The value of r lies between -1 and +1. A higher value of $r$ indicates a stronger linear relationship than a lower value of $r$. Further, the correlation coefficient is symmetric in the sense that $r$ between X and Y is the same as $r$ between Y and X.

Correlation coefficient however does not indicate any cause and effect relationship but only says that each variable in the pair moves in tandem with the other (r > 0) or opposing the other (r < 0).

When $r = +1$ it means a perfect positive linear (straight line) relationship and $r = -1$ is a perfect negative linear relationship. Again $r = 0$ implies that $X_j$ and $X_l$ are uncorrelated to mean that they lack a linear relationship between them. It however does not mean that these two variables are independent because Equation 1.6 assumes a linear relationship but there could be a quadratic or other non-linear relationship between them or there may not be any relationship at all.

The correlations measured from the data can be arranged in the form of a matrix called the *correlation matrix* given by

$$r = \begin{pmatrix} 1 & r_{12} & \cdots & r_{1k} \\ r_{21} & 1 & \cdots & r_{2k} \\ \cdots & \cdots & \cdots & \cdots \\ r_{k1} & r_{k2} & \cdots & 1 \end{pmatrix} \tag{1.7}$$

This is a symmetric matrix with lower and upper diagonal elements being equal. Such matrices have importance in multivariate analysis. Even though it is enough to display only the upper or lower elements above the diagonal, some software packages show the complete *(k × k)* correlation matrix.

The correlation coefficient has an important property whereby it remains unchanged when data is modified with a linear transformation (multiplying by a constant or adding a constant). In medical diagnosis some measurements are multiplied by a scaling factor like 100 or 1000 or $100^{-1}$ or $1000^{-1}$ for the purpose of interpretation. Sometimes the data is transformed to a percentage to overcome the effect of units of measurements but r remains unchanged.

Another important property of the correlation coefficient is that it is a *unit free* value or *pure number* to mean that with different units of measurement on $X_j$ and $X_l$, the formula Equation 1.6 produces a value without any units. This makes it possible to understand the strength of the relationship among several pairs of variables with different measurement units by just comparing the correlation coefficients.

**Coefficient of determination:**

The square of the correlation coefficient $r^2$ is called the *coefficient of determination*. The correlation coefficient (r) assumes that there is a linear relationship between X and Y. The larger the value of r, the stronger the linear relationship. Since r can be positive or negative, $r^2$ is always positive and is used to measure the amount of variation in one variable explained by the other variable. For instance, when r = 0.60, we get $r^2 = 0.36$ which means that only 36% of variation between the variables is explained by the correlation coefficient. There could be several reasons for such a low value of $r^2$ such as a wide scatter of points around the linear trend or a nonlinear relationship may be present which the correlation coefficient cannot detect.

Consider the following illustration.

**Illustration 1.2** Reconsider the data used in Illustration 1.1. We will examine the data to understand its properties and to know the inter-relationships among them. Let us consider a profile of measurements with four variables CHOL, TRIG, HDL and LDL.

From the data the following descriptive statistics can be obtained using SPSS → Analyze → Descriptives. One convention is to present the summary as mean ± S.D

| Profile | All cases (n = 64) | |
| Variable | Mean | Std. Dev. |
| --- | --- | --- |
| CHOL | 165.688 | 34.444 |
| TRIG | 134.672 | 75.142 |
| HDL | 36.281 | 5.335 |
| LDL | 102.488 | 28.781 |

The data contains a classification variable Group = 1 or 0. While processing the data we can attach the true labels instead of numeric codes. We can also view the above profile simultaneously for the two groups as shown in Table 1.2.

TABLE 1.2: Descriptive statistics for the profile variables

| Profile variables | Cases (n = 32) | | Controls (n = 32) | |
| | Mean | Std. Dev | Mean | Std. Dev |
| --- | --- | --- | --- | --- |
| CHOL | 164.156 | 28.356 | 167.219 | 40.027 |
| TRIG | 126.750 | 60.690 | 142.594 | 87.532 |
| HDL | 36.125 | 6.084 | 36.438 | 4.557 |
| LDL | 102.419 | 25.046 | 102.556 | 32.497 |

In the univariate analysis, we are interested in comparing the mean value of a parameter like HDL between the two groups to find whether the difference in the means could be considered as significant. This exercise is done for each variable, independent of others. In the multivariate context we compare the 'mean vector' between the two groups while in the univariate environment, the mean values are compared for one variable at a time.

With the availability of advanced computing tools, profiles can be *visualized* in two and three dimensions to understand what the data describes. For instance, the individual distributions of LDL and CHOL are shown by histograms separately in Figure 1.1a and Figure 1.1b which are univariate displays.

The joint distribution of LDL and CHOL is shown by a 3D-histogram in Figure 1.2. (Hint: In SPSS we use the commands; Graphs → Graphboard Templete Chooser → Select CHOL and LDL using Ctrl key → Press Ok. Manage colors).

Since there are only two variables, visualization is possible in three dimensions and with more than three variables it is not possible to visualize the pattern.

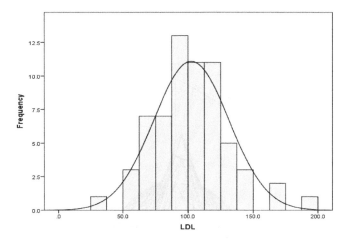

(a) Univariate distribution of LDL.

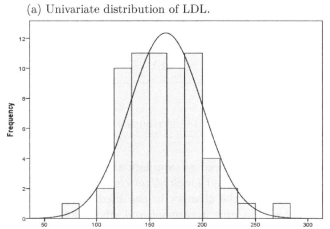

(b) Univariate distribution of CHOL.

FIGURE 1.1: Univariate distribution.

The maximum dimensions one can visualize are only three, viz., length, breadth and height/depth. Beyond three dimensions, data is understood by numbers only.

We will see in the following section that the four variables listed above are *correlated* to each. In other words, a change in the values of one variable causes a proportionate change in other variables. Therefore, when the profile variables are correlated to each, independent univariate analysis is not correct and multivariate analysis shall be used to compare the entire profile (all the four variables simultaneously) between the groups. If the difference is significant, then independent univariate comparison has to be made as a Post Hoc analysis.

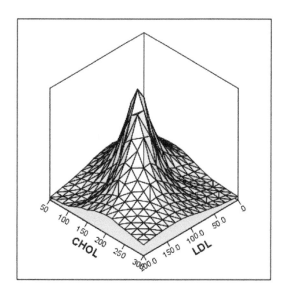

FIGURE 1.2: Bivariate distribution of LDL & CHOL.

In some situations, we need to compare the mean values of a variable at different time points (called *longitudinal* comparison) to observe changes over time. For instance, the changes in the lipid profile or hemogram will be compared at baseline, after treatment and after a follow-up. Multivariate comparison tests are used in this context.

In the next section we discuss a measure called *variance-covariance matrix* to express the spread among the components of a multivariate panel.

## 1.6   Covariance Matrix and Correlation Matrix

In addition to means and variances, the covariance structure of variables (within a profile) plays an important role in multivariate analysis. Covariance is a measure of the joint variation between two variables as defined in

Equation 1.4. For the 4-variable profile given in Illustration 1.2, the matrix of variances and covariances is obtained as shown below. Some software packages call this *covariance matrix* instead of variance-covariance matrix.

|       | CHOL     | TRIG     | HDL     | LDL     |
|-------|----------|----------|---------|---------|
| CHOL  | **1186.409** | 1007.213 | 93.042  | 888.936 |
| TRIG  | 1007.213 | **5646.319** | -36.224 | -68.891 |
| HDL   | 93.042   | -36.224  | **28.459**  | 70.785  |
| LDL   | 888.936  | -68.891  | 70.785  | **828.340** |

The variances are shown in boldface (along the diagonal) and the off-diagonal values indicate the covariances. The covariance terms represent the joint variation between pairs of variables. For instance the covariance between CHOL and LDL is 93.042. A higher value indicates more covariation.

For the end user, it is difficult to interpret the covariance because the units are expressed in *product of two natural units*. Instead of covariance, we can use a related measure called the *correlation coefficient* which is a pure number (free of any units). However, the mathematical treatment of several multivariate problems is based on the variance-covariance matrix itself.

The correlation matrix for the profile variables is shown below.

|       | CHOL  | TRIG   | HDL    | LDL    |
|-------|-------|--------|--------|--------|
| CHOL  | 1     | 0.389  | 0.506  | 0.897  |
| TRIG  | 0.389 | 1      | -0.090 | -0.032 |
| HDL   | 0.506 | -0.090 | 1      | 0.461  |
| LDL   | 0.897 | -0.032 | 0.461  | 1      |

The correlation coefficients on the diagonal line are all equal to 1 and all the upper diagonal values are identical to the lower diagonal elements. Since, by definition, the correlation between variables is symmetric, the lower diagonal values need not be shown. Some statistical packages like the Data Analysis Pak for MS-Excel show only the upper diagonal terms. Some MS-Excel Add-ins (for instance *Real Statistics Add-ins*) offer interesting Data Analysis Tools which can be added to MS-Excel.

More details on MS-Excel and SPSS for statistical analysis can be found in Sarma (2010).

It can be seen that CHOL has a strong positive relationship with LDL ($r = 0.897$), which means that when one variable increases, the other one also increases. Similarly, the correlation coefficient between LDL and CIMT is very low and negative.

Both the covariance matrix and the correlation matrix play a fundamental role in multivariate analysis.

The next section contains a discussion of methods for and the advantages of data visualization.

---

## 1.7   Data Visualization

Data visualization is an important feature in multivariate analysis which helps in understanding the relationships among pairs of variables. This can be done with a tool called a *matrix scatter plot* that simultaneously shows the plots of all pairs of related variables and also across groups of cases. Creation and interpretation of such plots are discussed in Illustration 1.1.

With the availability of computer software it is now easy to visualize the data on several variables simultaneously in a graph and understand the data pattern. Some important graphs used in data analysis and visualization are discussed below.

**Histogram:**

This is a graph used to understand the data pattern of a continuous variable with grouped distribution. The *count* or *frequency* of values that fall in an interval (called a *bin*) is plotted as a vertical rectangle with height proportional to the count and width proportionally equal to the interval. For normally distributed values the histogram will be *symmetric* with its peak neither high nor low. Data with many values lower than the mean will be *left skewed* and data with many values higher than the mean will be *right skewed*.

Consider the following illustration.

**Illustration 1.3**   Consider the data of Illustration 1.1. Let us examine the histogram of LDL.

The histogram can be constructed with the following SPSS options.

a) Graphs → Legacy Dialogs → Histogram.

b) Push LDL to variable.

c) Check display normal curve.

d) Press OK.

The chart is produced with minimum features. With a double click on the chart the required options can be inserted into the chart which looks like the one given in Figure 1.3. It is easy to see that the shape of the distribution is more or less symmetric. There are 10 patients in the LDL bin of 60-80. With a double click on the SPSS chart, we get options to change the bin width or the number of bins to show. With every change, the shape of the histogram changes automatically.

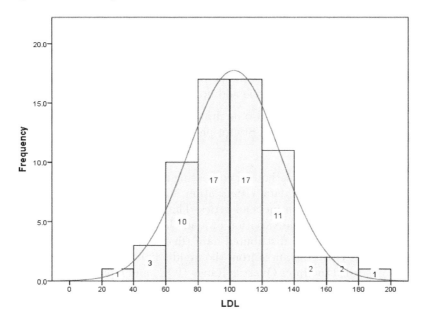

FIGURE 1.3: Histogram of LDL with normal distribution.

In well-distributed data, we expect a specific shape called the *normal distribution* as shown by the embedded curve on the vertical bars. In brief, the normal distribution has important properties like a) mean = median, b) symmetric around mean (skewness measure = 0) and c) neither flat nor peaked (kurtosis measure = 3).

No real life data shows a perfect normal shape but often exhibits a pattern close to it. The researcher can use either personal judgement or use another graphic procedure called the P-P plot to accept normality. Further, the shape of the histogram depends on the choice of the number of bins or bin width. (Double click on the histogram in the SPSS output and change the width to see the effect on the histogram!)

## Why normality?

Several statistical tools of inference are based on the assumption that the data can be explained by a theoretical model of normal distribution. In simple terms, normality indicates that about 95% of the individual value lies within 2 standard deviations on either side of the mean. Values outside this window can be considered as *abnormal* or simply *outliers*.

## Bar chart:

When the data is discrete like gender, case/control or satisfaction level, we do not draw a histogram but a *bar chart* is drawn. In a bar chart the individual bars are separated by a gap (to indicate that they are distinct categories!).

## Pie chart:

This chart is used to display the percentage of different cases as segments of a circle marked by different colors or line. The labels for each color could be either the actual number or the percentage. We use this chart only when the components within the circle add up to 100%.

All the above charts can also be drawn by using simple software like MS-Excel, MedCalc, Stata, R and Statgraphics.

## Box & Whisker plot:

This is a very useful method for comparing multivariate data. Tukey (1977) proposed this chart for data visualization and it is commonly used in exploratory analysis and business analytics. The concentration of data is shown as a vertical box and the variation around the median is shown by vertical lines. For symmetrically distributed data (line normal) the two edges of the box will be at equal distances from the middle line (median). The difference (Q3-Q1) is called the Inter Quartile Range (IQR) and represents the height of the box and holds 50% of the data. Typical box plots are shown in Figure 1.4.

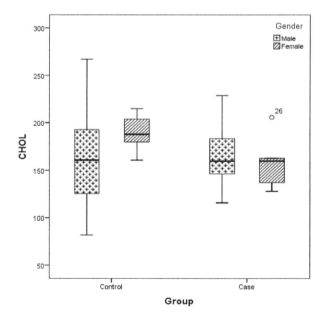

FIGURE 1.4: Box and Whisker plot for CHOL by gender and group.

The box plot is also used to identify the *outliers* which are values that are considered unusual or abnormal and defined as follows.

- Outliers are those values which are either a) above 3×IQR or more than the third quartile or b) below 3×IQR or less than the first quartile.

- Suspected outliers are either 1.5×IQR or more above the third quartile or 1.5×IQR or more below the first quartile.

There will be vertical lines called *whiskers* terminated by a crosshatch. Whiskers are drawn in two ways.

a) The ends of the whiskers are usually drawn from the top of the box to the *maximum* and from the bottom of the box to the *minimum*.

b) If either type of outlier is present, the whisker on the appropriate side is taken to 1.5×IQR from the quartile (the "inner fence") rather than the maximum or minimum. The individual outlying data points are displayed as unfilled circles (for suspected outliers) or filled circles (for outliers). (The "outer fence" is 3×IQR from the quartile.)

Lower Whisker = nearest data value larger than $(Q_1$-1.5×IQR) and Upper Whisker = nearest data value smaller than $(Q_3$+1.5×IQR).

When the data is normally distributed the median and the mean will be identical and further IQR = 1.35×$\sigma$ so that the whiskers are placed at a distance of 2.025 times or approximately at 2$\sigma$ from the median of the data.

In Figure 1.4 we observe that the CHOL for men in cases has an outlier labelled as 26. When the data has outliers, there are several methods of estimating mean, standard deviation and other parameters. This is known as *robust estimation*.

One practice is to exclude the top 5% and bottom 5% values and estimate the parameters (provided this does not suppress salient cases of data). For instance the *trimmed mean* is the mean value that is obtained after trimming top and bottom 5% extreme values and hence it is more reliable in the presence of outliers. The 'explore' option of SPSS gives this analysis.

**Scatter diagram:**

Another commonly used chart that helps visualization of correlated data is the scatter diagram. Let X and Y be two variables measured on an interval scale, like BMI, glucose level etc. Let there be n patients for whom data is available as pairs $(x_i, y_i)$ for i = 1, 2, $\cdots$, n. The plot of $y_i$ against $x_i$ marked as dots produces a shape similar to the scatter of a fluid on a surface and hence the name scatter diagram.

A scatter diagram indicates the nature of the relationship between the two variables as shown in Figure 1.5.

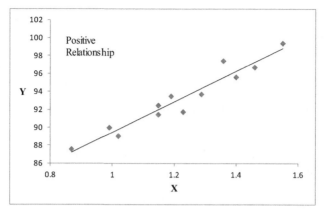

(a) Positive relationship.

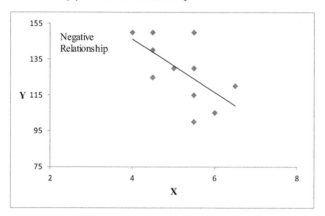

(b) Negative relationship.

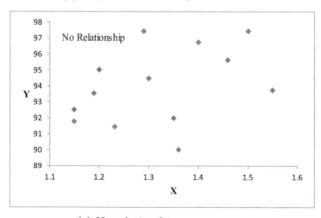

(c) No relationship.

FIGURE 1.5: Scatter diagram with direction of relationship.

Referring to Figure 1.5a it follows that Y increases whenever X increases and hence it is a case of positive relationship. In Figure 1.5b the relationship is in the opposite direction. Here, Y decreases as X increases and hence it is a case of a negative relationship. When no specific pattern appears as in Figure 1.5c, we observe that there is no clear linear relationship between X and Y.

Let us revisit Illustration 1.1 and examine the scatter diagrams of the profile variables with the help of SPSS. The chart is shown in Figure 1.6 which is known as a *scatter matrix*.

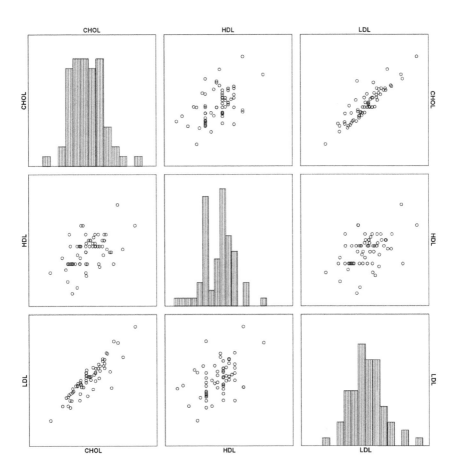

FIGURE 1.6: Scatter matrix for three variables.

It can be seen that a clear positive relationship exists between CHOL and LDL while other relationships are not clearly positive. This chart helps understand all possible correlations simultaneously. The histograms shown in the diagonal elements refer to the same variable for which the distribution is shown to understand the variation within that variable. This chart is produced with the SPSS option in the Graph board Template Chooser menu called *scatter plot matrix* (SPLOM).

In a different version of this chart, the diagonal elements in the matrix are left blank because the same variable is involved in the pair, for which no scatter exists.

When the scatter is too wide, it may indicate abnormal data and the scatter is vague without a trend. In such cases, removal of a few widely scattered points may lead to a recognizable pattern. One can do this type of exercise using MS-Excel for scatter charts directly.

A 3D scatter plot is another tool useful to understanding the multivariate scatter of three variables at a time. This can be worked out with SPSS from the Graph board Template Chooser but can be presented differently using R software and appears as shown in Figure 1.7. The following R-code is used to produce the 3D chart.

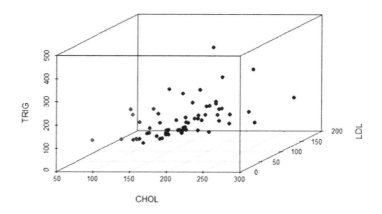

FIGURE 1.7: A 3D-Scatter chart using R.

**R Code:**
*library(MASS)*
# reading a file
*cimt.data<-read.csv(file.choose(),header = TRUE)*
*attach(cimt.data)*
# 3d Scatterplot

*library(scatterplot3d)*
*scatterplot3d(CHOL,LDL,TRIG, pch=16, highlight.3d=TRUE)*

In the following section we discuss a statistical model for multivariate normal distribution and its visualization.

---

## 1.8    The Multivariate Normal Distribution

Normal distribution plays a vital role in routine statistical analysis like testing of hypotheses. The characteristics of normal distribution for a single variable can be extrapolated to multivariate situations.

Let $\mathbf{X}$ be the vector of k-variables. Let $x_1, x_2, \ldots, x_k$ be the k-measurements made on one individual. Then we can characterize these measurements by a mathematical model called the *density function* given by

$$f(x_1, x_2, \ldots, x_k) = \frac{1}{\sqrt{(2\pi)^k \ |\ \mathbf{\Sigma}\ |}} e^{-\frac{1}{2}(\mathbf{X}-\mathbf{\mu})' \mathbf{\Sigma}^{-1} (\mathbf{X}-\mathbf{\mu})},$$

$$-\infty < x_1, x_2, \ldots, x_k < \infty \quad (1.8)$$

where $\mathbf{\mu}$ denotes the mean vector and $\mathbf{\Sigma}$ is the variance-covariance matrix given in Equation 1.5. This formula is similar to the univariate normal density when $k = 1$. In that case, we get only one mean $\mu$ and a single variance $\sigma^2$ so that Equation 1.8 reduces to

$$f(x) = \frac{1}{\sigma\sqrt{2\pi}} e^{-\frac{(x-\mu)^2}{2\sigma^2}}, -\infty < x < \infty \quad (1.9)$$

Several statistical procedures are based on the assumption that the data follows a multivariate normal distribution. It is not possible to visualize the distribution graphically when $k > 2$ since we can at most observe a 3-dimensional plot.

When $k = 2$ we get the case of *bivariate normal distribution* in which only two variable $X_1$ and $X_2$ are present and the parameters $\mathbf{\mu}$ and $\mathbf{\Sigma}$ of Equation 1.8 take the simple form $\mathbf{\mu} = \begin{bmatrix} \mu_1 \\ \mu_2 \end{bmatrix}$ where $\mu_1$ and $\mu_2$ are the means of $X_1$ and $X_2$ and $\mathbf{\Sigma} = \begin{bmatrix} \sigma_1^2 & \sigma_{12} \\ \sigma_{21} & \sigma_2^2 \end{bmatrix}$ where $\sigma_1^2$, $\sigma_2^2$ are the two variances and $\sigma_{12}$ is the covariance between $X_1$ and $X_2$.

If $\rho$ denotes the correlation coefficient between $X_1$ and $X_2$, then we can write the covariance as $\sigma_{12} = \rho\sigma_1\sigma_2$. Therefore in order to understand the bivariate normal distribution we need *five* parameters $(\mu_1, \mu_2, \sigma_1, \sigma_2$ and $\rho)$.

Thus, the bivariate normal distribution has a lengthy but interesting formula for the density function given by

$$f(x_1, x_2) = \frac{1}{2\pi\sigma_1\sigma_2\sqrt{1-\rho^2}} e^{-Q}, -\infty < x_1, x_2 < \infty \text{ where}$$

$$Q = \left[\frac{1}{2\sigma_1^2\sigma_2^2(1-\rho^2)}\right]\left[\left(\frac{x_1-\mu_1}{\sigma_1}\right)^2 + \left(\frac{x_2-\mu_2}{\sigma_2}\right)^2\right.$$

$$\left. - 2\rho\left(\frac{x_1-\mu_1}{\sigma_1}\right)\left(\frac{x_2-\mu_2}{\sigma_2}\right)\right] \quad (1.10)$$

Given the values of the five parameters, it is possible to plot the density function given in Equation 1.10 as a 3D plot. Visualization of bivariate normal density plot is given in Figure 1.8.

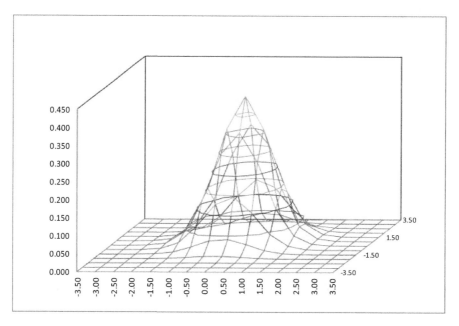

FIGURE 1.8: Simulated bivariate normal density.

For the case of more than two variables summary statistics like mean vectors and covariance matrices will be used in the analysis.

In the following section an outline is given of different applications of multivariate data. The importance of these tools in handling datasets with large numbers of variables and cases is also mentioned. Most of these applications are data-driven and need software to perform the calculations.

## 1.9   Some Interesting Applications of Multivariate Analysis

a) **Visual understanding of data:** We can use software to view and understand the distribution of several variables graphically. A histogram or a line chart can be used for this purpose. A multivariate scatter diagram is a special case of interest to understand the mutual relationships among variables.

b) **Hidden relationships**: There could be several direct and indirect relationships among variables which can be understood with the help of *partial correlations* and *canonical correlations*.

c) **Multivariate ANOVA (MANOVA)**: The multivariate analogue of Analysis of Variance (ANOVA) is known as *Multivariate ANOVA (MANOVA)*. Suppose we observe 3 response variables in 3 groups defined by treatment dose, say Placebo, 10mg and 20mg. For each group we get a vector (profile) of 3 means. Using MANOVA we can test for the significance of the difference among the 3 profiles (not individual variables) between groups. Once we confirm that the profiles differ significantly we employ one way ANOVA for each variable separately. If the profiles do not differ significantly, further analysis is not necessary.

d) **Repeated Measures Analysis (RMANOVA)**: Suppose we take the Fasting Blood Sugar (FBS) level of each of the 10 patients at two time points, viz., Fasting and PP. To compare the average glucose level at two time points, we use the *paired t-test*. Suppose the blood sugar is measured at 4 times, say at 9am, 11am, 1pm and 3pm. To compare these values we use a multivariate tool called *Repeated Measures Analysis of Variance*.

e) **Multiple regression** The joint effect of several explanatory variables on one response variable (Y) can be studied with the help of a multiple linear regression model. It is a cause and effect situation and we wish to study the joint as well as the marginal effect of input variables on the response. A commonly used method of regression is known as *stepwise regression*. It helps in identifying the most influential variables that affect the response.

f) **Principal components**: Sometimes psychological studies involve hundreds of variables representing the response to a question. It is difficult to handle a large number of variables in a regression model and poses a problem of *dimensionality*. Principal component analysis is a tool to reduce dimensionality. It so happens that among all the variables, some of them can be *combined* in a particular manner so that we produce a *mixture* of original variables. They indicate 'latent' features that are not directly observable from the study subjects. We can also extract several such mixtures so that they are uncorrelated! The number of such mixtures called *principal components* will be fewer than the number of original variables.

g) **Factor analysis**: Factor Analysis (FA) is another multivariate tool that works on the method of identifying the *latent* or *hidden* factors in the data. They are like the principal components and hence they are new entities called *factors*. For instance, *courtesy* could be a hidden factor for the success at a blood bank counter. It cannot be measured directly but can be expressed in different observable quantities like (1) warm receiving at the counter (yes/no), (2) throwing away the request form (yes/no), (3) careless answering (yes/no) or (4) too much talkative (yes/no) etc. These are four factors that might have been extracted from a large set of original variables.

h) **Classification problems:** In these problems we are interested in establishing a multiple variable model to predict a binary outcome like presence or absence of a disease or a health condition. By observing enough number of instances where the event is known, we develop a formula discriminant between the two groups. This is also called a *supervised learning* method by data scientists. When there are no predefined groups, the classification is called *unsupervised learning* and the technique used if cluster analysis.

We end this chapter with the observation that a holistic approach to understanding data is multivariate analysis.

---

## Summary

The conventional approach toward analysis in many studies is univariate. The statistical tests like t-test are aimed at comparing the mean values among treatment groups. This is usually done for each outcome variable at a time. The measurements made on a patient are usually correlated and this structure conveys more supportive information in decision making. Multivariate analysis is a class of analytical tools (mostly computer intensive) and they provide great insight into the problem.

We have highlighted the importance of vectors and matrices to represent multivariate data. The need for a correlation matrix and visual description of data helps to understand the inter-relationships among the variables of study.

# Do it yourself (Exercises)

**1.1** Consider the following data from 20 patients about three parameters denoted by X1, X2 and X3 on the Bone Mineral Density (BMD).

| S.No | Gender | X1 | X2 | X3 | S.No | Gender | X1 | X2 | X3 |
|------|--------|------|------|------|------|--------|------|------|------|
| 1 | Male | 0.933 | 0.820 | 0.645 | 11 | Male | 0.715 | 0.580 | 0.433 |
| 2 | Female | 0.889 | 0.703 | 0.591 | 12 | Female | 0.932 | 0.823 | 0.636 |
| 3 | Female | 0.937 | 0.819 | 0.695 | 13 | Male | 0.800 | 0.614 | 0.518 |
| 4 | Female | 0.874 | 0.733 | 0.587 | 14 | Female | 0.699 | 0.664 | 0.541 |
| 5 | Male | 0.953 | 0.824 | 0.688 | 15 | Male | 0.677 | 0.547 | 0.497 |
| 6 | Female | 0.671 | 0.591 | 0.434 | 16 | Female | 0.813 | 0.613 | 0.450 |
| 7 | Female | 0.914 | 0.714 | 0.609 | 17 | Male | 0.851 | 0.680 | 0.569 |
| 8 | Female | 0.883 | 0.839 | 0.646 | 18 | Male | 0.888 | 0.656 | 0.462 |
| 9 | Female | 0.749 | 0.667 | 0.591 | 19 | Female | 0.875 | 0.829 | 0.620 |
| 10 | Male | 0.875 | 0.887 | 0.795 | 20 | Male | 0.773 | 0.637 | 0.585 |

(a) Find out the mean BMD profile for male and female patients.

(b) Generate the covariance matrix and study the symmetry in the covariance terms.

(c) Obtain the correlation matrix and the matrix scatter plot for the profile without reference to gender.

**1.2** The following data refers to four important blood parameters, viz., Hemoglobin (Hgb), ESR, B12 and Ferritin obtained by a researcher in a hematological study from 20 patients.

(a) Construct the mean profile and covariance matrix among the variables.

(b) Draw a Box and Whisker plot for all the variables.

(c) Find the correlation matrix and identify which correlations are high (either positive or negative).

| S.No | Age | Hgb | ESR | B12 | Ferritin | S.No | Age | Hgb | ESR | B12 | Ferritin |
|------|-----|------|-----|-----|----------|------|-----|------|-----|-----|----------|
| 1 | 24 | 12.4 | 24 | 101 | 15.0 | 11 | 32 | 9.1 | 4 | 223 | 3.5 |
| 2 | 50 | 5.8 | 40 | 90 | 520.0 | 12 | 28 | 8.8 | 20 | 250 | 15.0 |
| 3 | 16 | 4.8 | 110 | 92 | 2.7 | 13 | 23 | 7.0 | 4 | 257 | 3.2 |
| 4 | 40 | 5.7 | 90 | 97 | 21.3 | 14 | 56 | 7.9 | 40 | 313 | 25.0 |
| 5 | 35 | 8.5 | 6 | 102 | 11.9 | 15 | 35 | 9.2 | 60 | 180 | 1650.0 |
| 6 | 69 | 7.5 | 120 | 108 | 103.0 | 16 | 44 | 10.2 | 60 | 88 | 11.2 |
| 7 | 23 | 3.9 | 120 | 115 | 141.0 | 17 | 46 | 13.7 | 10 | 181 | 930.0 |
| 8 | 28 | 12.0 | 30 | 144 | 90.0 | 18 | 48 | 10.0 | 40 | 252 | 30.0 |
| 9 | 39 | 10.0 | 70 | 155 | 271.0 | 19 | 44 | 9.0 | 30 | 162 | 44.0 |
| 10 | 14 | 6.8 | 10 | 159 | 164.0 | 20 | 22 | 6.3 | 40 | 284 | 105.0 |

*(Data courtesy: Dr. C. Chandar Sekhar, Department of Hematology,*
*Sri Venkateswara Institute of Medical Sciences (SVIMS), Tirupati.)*

**1.3** The following table represents in the covariance matrix among 5 variables.

| | Age | Hgb | ESR | B12 | Ferritin |
|---|------|------|------|------|------|
| Age | 184.193 | | | | |
| Hgb | 6.619 | 5.545 | | | |
| ESR | 85.729 | -35.074 | 1199.306 | | |
| B12 | 230.922 | 153.070 | 299.035 | 62449.460 | |
| Ferritin | 368.284 | 163.926 | 94.734 | -8393.780 | 136183.600 |

Obtain the correlation matrix by using the formula $r(x,y) = Cov(X,Y)/S_X * S_Y$.

**1.4** For the given data in Illustration 1.1, categorize age into meaningful groups and obtain the histogram CIMT for group = 1.

**1.5** Obtain the matrix-scatter plot of Hgb, ESR and B12 using SPSS from the data given in Exercise 1.2.

**1.6** A dot plot is another way of visualizing of data. MedCalc has some interesting graphs using dot plots. Use the data given in Table 1.1 in the Appendix and obtain the dot plot of (a) BMI and (b) CIMT groupwise and compare them with the box plot.

## Suggested Reading

1. Alvin C.Rencher, William F. Christensen. 2012. *Methods of Multivariate Analysis*. $3^{rd}$ ed. Brigham Young University: John Wiley & Sons.

2. Johnson, R.A., & Wichern, D.W. 2014. *Applied multivariate statistical analysis*, $6^{th}$ ed. Pearson New International Edition.

3. Ramzan, S., Zahid, F.M, Ramzan, S. 2013. Evaluating multivariate normality: A graphical approach. *Middle-East Journal of Scientific Research*. 13(2):254–263.

4. Anderson T.W. 2003. *An introduction to Multivariate Statistical Analysis*. $3^{rd}$ edition. New York: John Wiley.

5. Bhuyan K.C. 2005. *Multivariate Analysis and Its Applications*. India: New Central Book Agency.

6. Tukey, J.W. 1977. *Exploratory Data Analysis*. Addison-Wesley.

7. Crawley, M.J. 2012. *The R Book*: $2^{nd}$ ed. John Wiley & Sons.

8. Sarma K.V.S. 2010. *Statistics Made Simple-Do it yourself on PC*. $2^{nd}$ ed. Prentice Hall India.

# Chapter 2

# Comparison of Multivariate Means

| | | |
|---|---|---|
| 2.1 | Multivariate Comparison of Mean Vectors | 31 |
| 2.2 | One-sample Hotelling's $T^2$ Test | 33 |
| 2.3 | Confidence Intervals for Component Means | 37 |
| 2.4 | Two-Sample Hotelling's $T^2$ Test | 39 |
| 2.5 | Paired Comparison of Multivariate Mean Vectors | 43 |
| | Summary | 48 |
| | Do it yourself (Exercises) | 48 |
| | Suggested Reading | 50 |

Be approximately right rather than exactly wrong.

John Tukey (1915 – 2000)

## 2.1 Multivariate Comparison of Mean Vectors

In clinical studies it is often necessary to compare the mean vector of a panel of variables with a hypothetical mean vector to understand whether the observed mean vector is close to the hypothetical vector. Sometimes we may have to compare the panel mean of two or more independent groups of patients to understand the similarity among the groups. For instance, in cohort studies, one may wish to compare a panel of health indices among three or more categories of respondents like those based on the Socio Economic Status (SES).

The word 'vector' is used to indicate an array of correlated random variables or their summary values (like mean). If means are listed in the array, we call it the *mean vector*.

The univariate method of comparing the mean of each variable of the panel independently and reporting the p-value is not always correct. Since several variables are observed on one individual, the data is usually to be inter-correlated.

The correct approach to compare the mean vectors is to take into account the covariance structure within the data and develop a procedure that minimizes the error rate of false rejection. Multivariate statistical inference is based on this observation.

Let $\alpha$ be the type-I error rate for the comparison of the mean vectors. If there are k-variables in the profile and we make k-independent comparisons between the groups, the chance of making a correct decision will only be $\alpha' = 1 - (1 - \alpha)^k$, which is much less than the advertised error rate $\alpha$.

For instance with k $= 5$ and $\alpha = 0.05$, the ultimate error rate gets *inflated* to $\alpha' = 0.226$. It means about 23% of wrong rejections (even when the null hypothesis is true) are likely to take place against the promised 5% by way of univariate tests, which is a phenomenon known as *Rao's paradox* and more details can be had from Healy, M.J.R. (1969). Hummel and Sligo (1971) recommend performing a multivariate test followed by univariate t-tests.

There is a close relationship between testing of hypothesis and CI. Suppose we take $\alpha = 0.05$, then the 95% CI contains all plausible values for the true parameter $(\mu_0)$ specified under $H_0$. If $\mu_0$ is contained in the class interval we accept the hypothesis with 95% confidence; else we reject the hypothesis.

The Hotelling's $\mathbf{T^2}$ test provides a procedure to draw inferences on mean vectors as mentioned below.

a) Compare the sample mean vector with a hypothetical mean vector (claim of the researcher).

b) If the null hypothesis is not rejected at the $\alpha$ level, stop and conclude that the mean vectors do not differ significantly; *else*

c) Use 100 (1-$\alpha$)% CIs for each variable of the profile and identify which variable(s) contribute to rejection.

We now discuss the Hotelling's $\mathbf{T^2}$ test procedure for one sample problem and outline a computational procedure to perform the test and to interpret the findings. This will be followed by a two-sample procedure.

## 2.2 One-sample Hotelling's $T^2$ Test

Consider a profile $\mathbf{X}$ having k-variables $X_1, X_2, \cdots, X_k$ and assume that all are measured on a *continuous scale*. Assume that $\mathbf{X}$ follows a multivariate normal distribution with mean vector $\mu$ and covariance matrix $\Sigma$. Let the data be available from n-individuals. The sample mean vector $\overline{\mathbf{X}}$ defined in Equation 1.1 and the covariance matrix S defined in Equation 1.5 can be computed with the help of software like MS-Excel or SPSS.

We wish to test the hypothesis $\mathbf{H_0}$: $\mu = \mu_0$ versus $\mathbf{H_1}$: $\mu \neq \mu_0$, where $\mu_0$ is the hypothetical vector of k means defined as

$$\mu_0 = \begin{pmatrix} \mu_{01} \\ \mu_{02} \\ \vdots \\ \mu_{0k} \end{pmatrix}$$

The multivariate equivalent of the univariate t-statistic is given by the statistic

$$\mathbf{Z}^2 = n(\overline{\mathbf{X}} - \mu_0)'\Sigma^{-1}(\overline{\mathbf{X}} - \mu_0) \tag{2.1}$$

This statistic follows what is known as *Chi-square*($\chi^2$) distribution with k-degrees of freedom, under the assumption that $\Sigma$ is known. The decision rule is to reject $H_0$ if $\mathbf{Z}^2 > \chi^2_{k,\alpha}$ or if the p-value of the test is less than $\alpha$.

In general $\sigma$ is unknown and in its place $\mathbf{S}$ is used. Then $\mathbf{Z}^2$ reduces to the Hotelling's $\mathbf{T}^2$ statistic (see Anderson (2003)), given by

$$\mathbf{T}^2 = n(\overline{\mathbf{X}} - \mu_0)'\mathbf{S}^{-1}(\overline{\mathbf{X}} - \mu_0) \tag{2.2}$$

The $\mathbf{T}^2$ statistic is related to the well-known F-statistic by the relation $\mathbf{T}^2 = \dfrac{(n-1)k}{(n-k)} F_{k,n-k}$.

The critical value of $\mathbf{T}^2$ test can be easily found from F distribution tables. The p-value of the Hotelling's test can be got with the help of an MS-Excel function.

Standard statistical software like SPSS does not have a function to compute $\mathbf{T}^2$ directly. MS-Excel also has no direct function in the standard $f_x$ list but a few Add-ins of MS-Excel offer array formulas to compute $\mathbf{T}^2$.

For instance, the *Real Statistics Add-in* of MS-Excel has a direct function to compute the $\mathbf{T}^2$ statistic (source: http//www.real-statistics.com). It also contains a class of procedures to handle one-sample and two-sample $\mathbf{T}^2$ tests directly from raw data. Some interesting matrix operations are also available in it.

Suppose the Hotelling's test rejects the null hypothesis, then the *post-hoc analysis* requires finding out which of the k-components of the profile vector contributed to the rejection of the null hypothesis.

For the $i^{th}$ mean, $\mu_i$ the $100(1-\alpha)\%$ CI (often known as the $\boldsymbol{T^2}$ *interval* or simultaneous CIs) is given by the relation

$$\left[\bar{x}_i - \sqrt{\frac{k(n-1)}{(n-k)}}F_{k,(n-k),\alpha}\left(\frac{S_{ii}}{n}\right)\right] \leqslant \mu_i \leqslant \left[\bar{x}_i + \sqrt{\frac{k(n-1)}{(n-k)}}F_{k,(n-k),\alpha}\left(\frac{S_{ii}}{n}\right)\right]$$

where $S_{ii}$ is the variance of the $i^{th}$ variable and $F_{k,(n-k),\alpha}$ denotes the $(1-\alpha)\%$ critical value on the F distribution with $(k, n-k)$ degrees of freedom.

We also write this as an interval

$$\left[\bar{x}_i - \theta\sqrt{\frac{S_{ii}}{n}}, \ \bar{x}_i + \theta\sqrt{\frac{S_{ii}}{n}}\right] \tag{2.3}$$

where $\theta = \sqrt{\frac{k(n-1)}{(n-k)}}F_{k,(n-k),\alpha}$ is a constant.

If the hypothetical mean $\mu_i$ of the $i^{th}$ component lies outside this interval, we say that this variable contributes to rejection of the hypothesis and the difference between the observed and hypothetical means is significant.

Consider the following illustration.

**Illustration 2.1** In a cardiology study, data is obtained on various parameters from 50 patients with Metabolic Syndrome (MetS) and 30 patients without MetS. We call this *MetS data* for further reference. A description of variables, parameters and codes is given below.

| Variable | Parameter | Description |
|---|---|---|
| V1 | Group | With out MetS = 1, With MetS = 2 |
| V2 | Age | Age of the patient in years |
| V3 | Gender | 0 = Male, 1 = Female |
| V4 | Height | Height (inches) |
| V5 | Weight | Weight (Kg) |
| V6 | BMI | Body Mass Index $(Kg/m^2)$ |
| V7 | WC | Waist Circumference (cm) |
| V8 | HTN | Hypertension (0 = No, 1 = Yes) |
| V9 | DM | Diabetes Mellites (0 = No, 1 = Yes, 2 = Pre-diabetic) |
| V10 | TGL | Triglycerides (mg/dl) |
| V11 | HDL | High-density lipoproteins cholesterol (mg/dl) |
| V12 | LVMPI | Left Ventricle Myocardial Performance Index |

Table 2.1 shows a portion of data with 20 records but the analysis and discussion is carried out on 40 records of the dataset.

The profile variables are Weight, BMI, WC and HDL. The researcher claims that the mean values would be Weight = 75, BMI = 30, WC = 95 and HDL = 40.

TABLE 2.1: MetS data with a sample of 20 records

| S.No | V1 | V2 | V3 | V4 | V5 | V6 | V7 | V8 | V9 | V10 | V11 | V12 |
|------|----|----|----|------|------|------|-----|----|----|-----|-----|-------|
| 1 | 1 | 35 | 1 | 175.0 | 76.0 | 24.8 | 93 | 0 | 0 | 101 | 38 | 0.480 |
| 2 | 1 | 36 | 0 | 160.0 | 65.0 | 25.8 | 91 | 0 | 0 | 130 | 32 | 0.550 |
| 3 | 1 | 40 | 1 | 168.0 | 72.0 | 25.5 | 85 | 0 | 0 | 82 | 32 | 0.420 |
| 4 | 1 | 30 | 0 | 158.0 | 59.0 | 23.6 | 75 | 0 | 0 | 137 | 33 | 0.590 |
| 5 | 1 | 45 | 0 | 156.0 | 63.0 | 25.9 | 75 | 0 | 0 | 171 | 38 | 0.470 |
| 6 | 1 | 54 | 1 | 164.0 | 60.0 | 22.3 | 75 | 0 | 0 | 97 | 38 | 0.410 |
| 7 | 1 | 43 | 0 | 152.0 | 39.0 | 16.9 | 68 | 0 | 0 | 130 | 50 | 0.450 |
| 8 | 1 | 35 | 1 | 176.0 | 73.0 | 23.6 | 95 | 0 | 0 | 70 | 40 | 0.430 |
| 9 | 1 | 32 | 1 | 165.0 | 52.0 | 19.2 | 76 | 0 | 0 | 96 | 42 | 0.410 |
| 10 | 1 | 32 | 1 | 178.0 | 72.0 | 22.7 | 88 | 0 | 0 | 102 | 47 | 0.380 |
| 11 | 2 | 39 | 0 | 152.0 | 67.0 | 29.0 | 94 | 0 | 1 | 205 | 32 | 0.730 |
| 12 | 2 | 39 | 1 | 182.0 | 94.0 | 28.3 | 111 | 0 | 1 | 150 | 32 | 0.660 |
| 13 | 2 | 47 | 1 | 167.0 | 83.0 | 30.0 | 98 | 0 | 1 | 186 | 32 | 0.600 |
| 14 | 2 | 37 | 0 | 153.0 | 62.0 | 26.5 | 85 | 0 | 1 | 183 | 35 | 0.520 |
| 15 | 2 | 36 | 0 | 160.0 | 68.0 | 23.4 | 93 | 0 | 1 | 150 | 37 | 0.600 |
| 16 | 2 | 41 | 1 | 172.0 | 67.0 | 22.6 | 94 | 0 | 1 | 91 | 36 | 0.750 |
| 17 | 2 | 42 | 0 | 162.5 | 73.0 | 27.6 | 100 | 0 | 1 | 115 | 42 | 0.500 |
| 18 | 2 | 43 | 1 | 162.5 | 67.5 | 25.6 | 106 | 1 | 1 | 312 | 31 | 0.730 |
| 19 | 2 | 31 | 1 | 160.0 | 76.0 | 29.6 | 100 | 0 | 0 | 180 | 35 | 0.610 |
| 20 | 2 | 45 | 1 | 167.0 | 60.0 | 21.5 | 90 | 1 | 1 | 233 | 35 | 0.520 |

*(Data courtesy: Dr D. Rajasekhar, Department of Cardiology, Sri Venkateswara Institute of Medical Sciences (SVIMS), Tirupati.)*

The profile vector is $\mathbf{X} = \begin{bmatrix} \text{Weight} \\ \text{BMI} \\ \text{WC} \\ \text{HDL} \end{bmatrix}$ and $\boldsymbol{\mu_0} = \begin{bmatrix} 75.0 \\ 29.0 \\ 95.0 \\ 40.0 \end{bmatrix}$ will be the hypothesized mean vector.

We wish to test the hypothesis where the sample profile represents a population claimed by the researcher with mean vector $\boldsymbol{\mu_0}$.

**Analysis:**

The following stepwise procedure can be implemented in MS-Excel (2010 version).

*Step-1:* Enter the data in an MS-Excel sheet with column headings in the first row.

*Step-2:* Find the mean of each variable and store it as a column with heading "means". This gives the mean vector. (Hint: Select the cells B42 to E42 and click AutoSum → Average. Then select a different area at the top of the sheet and select 4 'blank' cells vertically. Type the function =TRANSPOSE(B42 : E42) and press Control+Shift+Enter.)

*Step-3:* Enter the hypothetical mean vector ($\mu_0$) in cells G3 to G6.

*Step-4:* Click on Add-ins → Real Statistics → Data Analysis Tools → Multivar → Hotelling T-Square and Click.

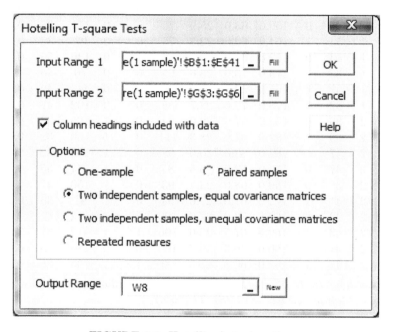

FIGURE 2.1: Hotelling's test options.

This gives an option window as shown in Figure 2.1. Choose the option One-sample.

*Step-5:* For the input range1, select the entire data from B1 to E41 (including headings) and for the input range2, select the hypothetical mean vector along with headings (G3:G6).

*Step-6:* Fix a cell to indicate the output range (for display of results) and press OK.

This gives $\mathbf{T}^2 = 3.916$ and $F = 0.9036$ with $p = 0.4721$. The hypothesis is accepted since $p > 0.05$. Hence, there is no significant difference between the sample mean vector and the hypothesized mean vector.

Suppose the hypothetical mean vector is changed to $\boldsymbol{\mu_0} = \begin{bmatrix} 75.0 \\ 30.0 \\ 90.0 \\ 90.0 \end{bmatrix}$.

From the worksheet of the Hotelling's $\mathbf{T}^2$ test, it is enough to make changes in the $\boldsymbol{\mu_0}$ and the results get automatically updated (one need not run the above steps again!).

This gives $\mathbf{T}^2 = 31.027$ with p-value $= 0.000238$ and the hypothesis is rejected, indicating a significant difference between the observed and claimed mean vectors. Therefore the difference between the sample mean vector and the hypothesized mean vector is significant at the 0.05 level.

**Remarks-1:**

When the profile has a mean vector which does not differ from a hypothetical vector, then $\mathbf{T}^2$ test is conclusive. Otherwise, one needs to find out which component(s) contributed to the rejection.

In the following section we discuss the use of CI in interpreting the results.

---

## 2.3 Confidence Intervals for Component Means

The next job is to generate a CI for the mean of each component and find which component is contributing to the rejection of the hypothesis.

The multiplier in Equation 2.3 is often called the *confidence coefficient* and the value of $F_{k,(n-k),\alpha}$ can be calculated using the MS-Excel function FINV()(see Sarma (2010)). This is called *critical value* of F, denoted by $F_{cri}$. For this example, we get $F_{cri} = 2.6335$ (using FINV($L$11,$H$5,$H$6)).

With simple calculations in the MS-Excel sheet we can build the intervals as shown in Figure 2.2. Some of the rows in the MS-Excel sheet are hidden for visibility of intermediate results.

Since the hypothetical mean of 90.0 for the variable WC falls outside the CI, we infer that WC differs significantly from the belief of the researcher. The other variables can be considered to agree with the hypothetical means.

When the number of comparisons is small, a better method of building CIs, instead of the $\mathbf{T}^2$ intervals is based on *family-wise error rate* known as *Bonferroni intervals* basing on the Student's t-test.

| Hotelling T-square Test | | | | | | | |
| One-sample test | | | | | | | |
| | | | | | | | |
| T2 | 30.8882693 | | | | | | |
| df1 | 4 | | | | | | |
| df2 | 36 | | | | | | |
| F | 7.12806216 | | | | | | |
| p-value | 0.00024659 | | | | | | |
| | | | | | | | |
| | | Computation of Simultaneous Confidence intervals | | | | | |
| n | 40 | COUNT(B2:B41) | | | alpha | 0.05 | |
| k | 4 | COUNTA(B1:E1) | | | | | |
| | | | | | 95% confidence interval | | |
| Variable | Hyp.Mean | Sample Mean | Variance | Conf.coeff | Lower | Upper | Significance |
| Weight | 75 | 76.77 | 290.237 | 4.9509 | 71.817 | 81.718 | Not Sig |
| BMI | 30 | 28.77 | 38.983 | 1.8145 | 26.951 | 30.579 | Not Sig |
| WC | 90 | 97.18 | 310.866 | 5.1239 | 92.051 | 102.299 | Sig |
| HDL | 40 | 39.80 | 53.087 | 2.1174 | 37.683 | 41.917 | Not Sig |
| | | | | | | | |
| | | Computation of Bonferroni intervals | | | | | |
| n | 40 | COUNT(B2:B41) | | | alpha | 0.05 | |
| k | 4 | COUNTA(B1:E1) | | | alpha-dash | 0.0125 | |
| | | | | | 95% confidence interval | | |
| Variable | Hyp.Mean | Sample Mean | Variance | Conf.coeff | Lower | Upper | Significance |
| Weight | 75 | 76.80 | 290.626 | 7.7921 | 69.008 | 84.592 | Not Sig |
| BMI | 30 | 28.77 | 38.983 | 2.8538 | 25.911 | 31.619 | Not Sig |
| WC | 90 | 97.18 | 310.866 | 8.0588 | 89.116 | 105.234 | Not Sig |
| HDL | 40 | 39.80 | 53.087 | 3.3303 | 36.470 | 43.130 | Not Sig |

FIGURE 2.2: MS-Excel worksheet to compute post-hoc calculations.

If there are k-means to compare then these intervals are given by

$$\left[ \bar{x}_i - t_{n-1,\alpha/2k}\sqrt{\left(\frac{S_{ii}}{n}\right)}, \bar{x}_i + t_{n-1,\alpha/2k}\sqrt{\left(\frac{S_{ii}}{n}\right)} \right] \quad (2.4)$$

where $t_{n-1,\alpha/2k}$ denotes the critical value ($t_{cri}$) on the t-distribution with (n-1) degrees of freedom and the error rate is re-defined as $\alpha_I = \dfrac{\alpha}{k}$.

If $\alpha = 0.05$ and k = 4 we get $\dfrac{\alpha}{k} = 0.05/4 = 0.0125$ and the critical value can be found from the TINV() function of Excel. In the above example we get $t_{cri} = 2.8907$ (using TINV(($L$22/2),($H$21-1))).

It may be seen that in the case of simultaneous CIs using Equation 2.3, the critical value was $F_{cri} = 2.6335$. The width of the CI for each of the variables depends on this critical value. Since $t_{cri} > F_{cri}$, the Bonferroni intervals are wider than the $\mathbf{T^2}$ intervals as shown in Table 2.2.

TABLE 2.2: Confidence limits for the components of the mean vector

| Profile variable | Hypothetical mean | Simultaneous ($\mathbf{T^2}$) limits | | | Bonferroni limits | | |
|---|---|---|---|---|---|---|---|
| | | Lower | Upper | Sig. | Lower | Upper | Sig. |
| Weight | 75 | 71.817 | 81.718 | No | 69.008 | 84.592 | No |
| BMI | 30 | 26.951 | 30.579 | No | 25.911 | 31.619 | No |
| WC | 90 | 92.051 | 102.299 | Yes | 89.116 | 105.234 | No |
| HDL | 40 | 37.683 | 41.917 | No | 36.47 | 43.13 | No |

While the $\mathbf{T^2}$ limits show WC as not a probable mean to believe as hypothetical, the Bonferroni limits show all the means acceptable under the hypothesis. A comparison of various methods of building CIs is based on statistical arguments and not discussed in this context. More details can be found in Johnson and Wichern (2002).

In the following section we discuss the $\mathbf{T^2}$ test for comparing the mean vectors of a panel between two independent groups (of records / individual) which is a two-sample test.

## 2.4 Two-Sample Hotelling's $\mathrm{T}^2$ Test

This test is used to compare the mean vectors of a profile between two independent groups like cases and controls, male and female etc. Let $\boldsymbol{\mu_1}$ and $\boldsymbol{\mu_2}$ be the mean vectors of the profile in group-1 and group-2 respectively. We wish to test the hypothesis $\mathbf{H_0}$: $\boldsymbol{\mu_1} = \boldsymbol{\mu_2}$ against $\mathbf{H_1}$: $\boldsymbol{\mu_1} \neq \boldsymbol{\mu_2}$ based on samples of size $n_1$ and $n_2$ from the two groups respectively.

We assume that $\boldsymbol{\Sigma_1} = \boldsymbol{\Sigma_2} = \boldsymbol{\Sigma}$ which means that the covariance matrices are equal in the two groups and further $\boldsymbol{\Sigma}$ is assumed to be known. In practice $\boldsymbol{\Sigma}$ is estimated as the pooled covariance matrix. The equality of two covariance matrices $\boldsymbol{\Sigma_1}$ and $\boldsymbol{\Sigma_2}$ is tested by the Box's M-test (1949), which is available in several software packages.

The Box's M statistic follows F distribution and the equality of variance is rejected if the p-value is very small. The Box test was originally designed for comparing several ($> 2$) covariance matrices and used in a more general context called MANOVA, discussed in Chapter 3.

If $\mathbf{S_1}$ and $\mathbf{S_2}$ denote the sample covariance matrices of the profile variables in group-1 and group-2 respectively then the pooled covariance matrix $\mathbf{S_{pl}}$ is given by

$$\mathbf{S_{pl}} = \left( \frac{n_1 - 1}{n_1 + n_2 - 2} \right) \mathbf{S_1} + \left( \frac{n_2 - 1}{n_1 + n_2 - 2} \right) \mathbf{S_2} \qquad (2.5)$$

The $\mathbf{T^2}$ statistic for testing $H_0$ is

$$\mathbf{T^2} = [\overline{\mathbf{X}}_1 - \overline{\mathbf{X}}_2]' \left[ \left( \frac{1}{n_1} + \frac{1}{n_2} \right) \mathbf{S}_{pl} \right]^{-1} [\overline{\mathbf{X}}_1 - \overline{\mathbf{X}}_2] \qquad (2.6)$$

where $\overline{\mathbf{X}}_1$ and $\overline{\mathbf{X}}_2$ denote the mean vectors in the two groups respectively.

For a given sample, after finding $\mathbf{T^2}$, we find the test value as

$$F_{k,n_1+n_2-k-1} = \frac{n_1 + n_2 - k - 1}{(n_1 + n_2 - 2)k} T^2$$

The critical value $(T^2_{cri})$ can be found from the F distribution with $(k,(n_1+n_2-k-1))$ degrees of freedom at the desired level $\alpha$.

The p-value of the test can be obtained from MS-Excel functions. If it is less than $\alpha$, the hypothesis of equal mean vectors can be rejected.

Now in order to find out which components of the profile mean vectors differ significantly, we have to examine the simultaneous CIs and check whether any interval contains zero (the hypothetical difference).

For the $i^{th}$ component, the simultaneous CIs are given by

$$\left[ (\overline{x}_{1i} - \overline{x}_{2i}) - \eta \sqrt{ \left( \frac{1}{n_1} + \frac{1}{n_2} \right) S_{ii}}, \; (\overline{x}_{1i} - \overline{x}_{2i}) + \eta \sqrt{ \left( \frac{1}{n_1} + \frac{1}{n_2} \right) S_{ii}} \right] \quad (2.7)$$

where $S_{ii}$ denotes the variance of the $i^{th}$ component in the *pooled covariance matrix* and

$\eta = \sqrt{ \dfrac{k(n_1 + n_2 - 2)}{(n_1 + n_2 - k - 1)} F_{k,(n_1+n_2-k-1),\alpha}}$ is the confidence coefficient which

is a constant for fixed values of $n_1, n_2, \alpha$ and $k$.

### Computational guidance:

The two sample $\mathbf{T^2}$ test can be done in MS-Excel by choosing the following options from the menu.

> *Add-ins* → *Real Statistics* → *Data Analysis Tools* → *Multivar* → *Hotelling T-Square* → *Two independent samples, equal covariance matrices* → *Press OK.*

MS-Excel requires the data of the two independent groups to be stored in two separate locations for analysis. The following is the procedure.

The null hypothesis is that there is no difference in the mean vectors. Hence the hypothetical vector has all zero values.

*Step-1:* Under Array1, select the data on the required variables including headings corresponding to group-1.

*Step-2:* Under Array2, select the data on the required variables including headings corresponding to group-2.

*Step-3:* Select the option 'Two independent samples' with 'equal covariance matrices'.

*Step-4:* Select a suitable cell for output range and press OK.

This directly gives $\mathbf{T^2}$ value and its significance (p-value).

When the hypothesis is rejected, further investigation requires the CIs which are not readily available after the $\mathbf{T^2}$ test from the Real Statistics Add-ins.

However, with simple calculations we can do this exercise as discussed in the following illustration.

**Illustration 2.2**  Let us reconsider the complete data used in Illustration 2.1 with the 4 profile variables, Weight, BMI, WC and HDL along with the grouping variable with codes 1, 2.

Putting this data into the MS-Excel sheet with the two groups stored in separate locations, we find the following intermediate values, before performing the $\mathbf{T^2}$ test.

a) Sample means and standard deviations

| Variable | Group-1 | | Group-2 | | Mean Difference | |
|---|---|---|---|---|---|---|
| | Mean | SD | Mean | SD | Sample | Hypothetical |
| Weight | 67.033 | 12.789 | 76.248 | 13.629 | 9.218 | 0 |
| BMI | 24.753 | 4.025 | 29.0460 | 4.813 | 4.293 | 0 |
| WC | 87.100 | 10.466 | 99.420 | 13.632 | 12.320 | 0 |
| HDL | 40.700 | 7.715 | 36.920 | 4.985 | -3.780 | 0 |

b) Sample covariance matrices

| $\mathbf{S_1}$ | | | | $\mathbf{S_2}$ | | | |
|---|---|---|---|---|---|---|---|
| 163.551 | 41.960 | 114.410 | -34.197 | 185.661 | 56.115 | 120.075 | -2.051 |
| 41.960 | 16.203 | 27.470 | -9.477 | 56.115 | 23.169 | 40.713 | -0.560 |
| 114.410 | 27.470 | 109.541 | -8.969 | 120.075 | 40.713 | 185.840 | -8.986 |
| -34.197 | -9.477 | -8.969 | 59.528 | -2.051 | -0.560 | -8.986 | 24.851 |

c) Pooled sample covariance matrix ($\mathbf{S_{pl}}$)

$$
\begin{array}{|rrrr|}
\hline
177.441 & 50.852 & 117.969 & -14.003 \\
50.852 & 20.579 & 35.789 & -3.875 \\
117.969 & 35.789 & 157.473 & -8.98 \\
-14.003 & -3.875 & -8.98 & 37.743 \\
\hline
\end{array}
$$

Making use of (i) the difference in the mean vectors and (ii) the matrix ($\mathbf{S_{pl}}$), the value of $\mathbf{T^2}$ is calculated with the following matrix multiplication using the array formula

=MMULT(TRANSPOSE(U5:U8),MMULT((1/(1/P5+1/S5))*

MINVERSE(N20:Q23),U5:U8)).

After typing the formula, we have to press Control+Shift+Enter. This gives $\mathbf{T^2} = 30.0790$.

The F value and the p-value are calculated using the MS-Excel functions as shown in Figure 2.3.

| | | Group-1 | | | | Group-2 | | | |
|---|---|---|---|---|---|---|---|---|---|
| | Mean | SD | n1 | | Mean | SD | n2 | Mean diff | |
| Weight | 67.033 | 12.789 | 30 | | 76.248 | 13.626 | 50 | 9.215 | |
| BMI | 24.753 | 4.025 | 30 | | 29.046 | 4.813 | 50 | 4.293 | |
| WC | 87.100 | 10.466 | 30 | | 99.420 | 13.632 | 50 | 12.320 | |
| HDL | 40.700 | 7.715 | 30 | | 36.920 | 4.985 | 50 | -3.780 | |

Group-1
Sample Covariance matrix

Group-2
Sample Covariance matrix

| 163.551 | 41.960 | 114.410 | -34.197 | | 185.661 | 56.115 | 120.075 | -2.051 |
|---|---|---|---|---|---|---|---|---|
| 41.960 | 16.203 | 27.470 | -9.477 | | 56.115 | 23.169 | 40.713 | -0.560 |
| 114.410 | 27.470 | 109.541 | -8.969 | | 120.075 | 40.713 | 185.840 | -8.986 |
| -34.197 | -9.477 | -8.969 | 59.528 | | -2.051 | -0.560 | -8.986 | 24.851 |

Pooled Covariance matrix

T-SQUARE
30.079

| 177.441 | 50.852 | 117.969 | -14.003 | | F | |
| 50.852 | 20.579 | 35.789 | -3.875 | | | 7.2305 |
| 117.969 | 35.789 | 157.473 | -8.980 | | p-val | |
| -14.003 | -3.875 | -8.980 | 37.743 | | | 0.000056 |

| | Alpha | 0.05 | | | | diff | Lower | Upper | Remark |
|---|---|---|---|---|---|---|---|---|---|
| | k | 4 | | Weight | | 9.2147 | -0.6935 | 19.1229 | Not Sig |
| | n1+n2-k-1 | 75 | | BMI | | 4.2927 | 0.9184 | 7.6669 | Sig |
| | | | | WC | | 12.3200 | 2.9859 | 21.6541 | Sig |
| | Simultaneous Confidence intervals | | | HDL | | -3.7800 | -8.3497 | 0.7897 | Not Sig |
| | F-cri (0.05) | 2.493696 | | | | | | | |
| | Factor | 4.16 | | | | | | | |
| | c | 3.220835 | | | | | | | |

FIGURE 2.3: MS-Excel worksheet for two-sample $T^2$ test.

The output gives $\mathbf{T^2} = 30.079$ with F = 7.2305 and p = 0.000056 and hence the hypothesis of equal mean vectors is rejected at the $\alpha = 0.05$ level (by default). It means there is a significant difference between the mean profiles of group-1 and group-2.

The simultaneous CIs are obtained by using the expression Equation 2.7 and calculations are shown in Figure 2.3. We notice that Weight and HDL contribute to the rejection of the hypothesis. Hence with 95% confidence we may conclude that the means of these two variables differ significantly (in the presence of other variables!).

If we use the MS-Excel Add-ins for Hotelling's two-sample test, we get essentially the same results but the we have to calculate the CIs separately.

**Remark-2:**

In the above evaluations, sample covariance matrices are obtained from *Real Statistics → Data Analysis Tools → Matrix operations → Sample covariance matrix*. This is a convenient method and the result can be pasted at the desired location.

**Remark-3:**

Hotelling's test is a multivariate test in which there should be at least two variables in the profile. If we attempt to use this method for a single variable in the profile, MS-Excel shows only error and no output is produced.

In the next section we discuss another application of the $\mathbf{T^2}$ test for comparing the mean vectors of two correlated groups, called *paired comparison test*.

---

## 2.5 Paired Comparison of Multivariate Mean Vectors

In clinical studies it is often necessary to compare a group of variables before and after giving a treatment. One convention is to call these two as *pre-* and *post-values*. In the univariate case this is done by using a *paired sample test*. In the univariate context let $(x_1,y_1)$, $(x_2,y_2)$, ..., $(x_n,y_n)$ denote the paired observations on the variable from each individual where $x_i$ and $y_i$ denote the measurement in the pre and post contexts.

We wish to test the hypothesis $H_0$: $\mu_x = \mu_y$ against $H_1$: $\mu_x \neq \mu_y$ where $\mu_x$ and $\mu_y$ denote the means of pre- and post-values, denoted by X and Y variables respectively. Since both measurements relate to the same individual, they are correlated and we cannot use the independent sample t-test.

Now define $d_i = (x_i - y_i) \; \forall \; i = 1, 2, \ldots, n$ and compute the mean difference $\bar{d} = \dfrac{1}{n} \sum\limits_{i=1}^{n} d_i$ and the variance of the difference $s_d^2 = \dfrac{1}{n-1} \sum\limits_{i=1}^{n} (d_i - \bar{d})^2$.

Testing $H_0: \mu_x = \mu_y$ is equivalent to testing $H_0: \mu_D = 0$ against $H_1: \mu_D \neq 0$, so that the test formula becomes $t = \dfrac{\sqrt{n}\bar{d}}{s_d}$, which follows Student's t-distribution with (n-1) degrees of freedom and the p-value of the test is compared with the level of significance $\alpha$. If the p-value is smaller than $\alpha$ we reject $H_0$ and consider that there is a significant difference between the pre and post means.

The method of using the difference $d$ reduces the problem to a one-sample t-test and hence the $100(1-\alpha)\%$ CIs for the true difference are given by

$$\left[ \bar{d} - t_{n-1} \frac{s_d}{\sqrt{n}} , \; \bar{d} + t_{n-1} \frac{s_d}{\sqrt{n}} \right] \tag{2.8}$$

SPSS has a module to perform this test with menu sequence Analyse $\rightarrow$ Compare means $\rightarrow$ Paired sample test. The data on X and Y will be stored in two separate columns. A similar procedure is available in MS-Excel Add-ins also.

Let us now extend the logic to a multivariate context.

**Hotelling's paired sample test for mean vectors:**

The univariate approach can be extended to the multivariate case where instead of a single variable we use a panel of variables $\mathbf{X}$ and $\mathbf{Y}$ for the pre- and post-treatment values. Each panel (vector) contains k-variables and the data setup can be represented as shown below with k = 4 variables.

For the $j^{\text{th}}$ variable, if we define the vector of differences $D_j = (X_j - Y_j)$ for

$j = 1, 2, \ldots, k$ we get new vectors $D_1, D_2, \ldots, D_k$ and the vector $\mathbf{D} = \begin{bmatrix} D_1 \\ D_2 \\ .. \\ D_k \end{bmatrix}$

is the new panel of variables which is assumed to follow multivariate normal distribution with mean vector $\mu_{\mathbf{D}}$ and variance covariance matrix $\Sigma_{\mathbf{D}}$.

We wish to test the hypothesis $\mathbf{H_0: \mu_x = \mu_y}$ against $\mathbf{H_1: \mu_x \neq \mu_y}$. Now the hypothesis $\mathbf{H_0: \mu_x = \mu_y}$ is equivalent to testing $\mathbf{H_0: \mu_d = 0}$ (Null vector) against $\mathbf{H_1: \mu_d \neq 0}$. This is similar to the one-sample Hotelling's $\mathbf{T^2}$ test.

Now define the mean vector and covariance matrix for $\mathbf{D}$ given as

$$\bar{\mathbf{D}} = \frac{1}{n} \sum_{j=1}^{n} D_j \text{ and } \mathbf{S_D} = \frac{1}{n-1} \sum_{j=1}^{k} (D_j - \bar{D})(D_j - \bar{D})'.$$

The test statistic is $\mathbf{T^2} = n \, \bar{\mathbf{D}}' \, \mathbf{S_D^{-1}} \, \bar{\mathbf{D}}$.

If the p-value of the test is smaller than $\alpha$, we reject $\mathbf{H_0}$ and compute the $100(1-\alpha)\%$ CIs for each of the k components of the panel by using

$$
\left[ \bar{d}_i - \sqrt{\frac{k(n-1)}{(n-k)} F_{k,(n-k),\alpha}} \left( \frac{s_{D_{ii}}}{n} \right), \right.
$$

$$
\left. \bar{d}_i + \sqrt{\frac{k(n-1)}{(n-k)} F_{k,(n-k),\alpha}} \left( \frac{s_{D_{ii}}}{n} \right) \right]
\tag{2.9}
$$

where $S_{D_{ii}}$ denotes the variance of the $i^{\text{th}}$ variable in matrix $\mathbf{S_D}$.

If any of the CIs does not contain zero, we conclude that the corresponding variable shows a significant difference. If the null hypothesis is not rejected we may skip computing the CIs.

Consider the following illustration.

**Illustration 2.3** The following is a portion of data obtained in a study on the effectiveness of a food supplement on endocrine parameters viz., T3, T4 and TSH. The supplement was administered to 90 individuals under three groups, viz., Group-1 (Control, n = 30), Group-2 (intervention-Spirulina Capsules, n = 30) and Group-3 (intervention-Spirulina Bar, n = 30).

Table 2.3 shows a sample of 15 records with values on these parameters before and after treatment supplementation. The values of the parameters before and after treatment are marked PRE and PO respectively.

TABLE 2.3: Thyroid panel values before and after treatment

| S.No | Group | T3_PRE | T4_PRE | TSH_PRE | T3_PO | T4_PO | TSH_PO |
|------|-------|--------|--------|---------|-------|-------|--------|
| 1 | 1 | 89 | 10.0 | 0.41 | 90 | 9.8 | 0.40 |
| 2 | 1 | 103 | 9.2 | 3.80 | 100 | 9.3 | 3.70 |
| 3 | 1 | 113 | 6.9 | 2.50 | 112 | 6.8 | 2.60 |
| 4 | 1 | 78 | 10.5 | 1.30 | 80 | 10.1 | 1.40 |
| 5 | 1 | 63 | 8.5 | 1.50 | 65 | 9.0 | 1.60 |
| 6 | 2 | 100 | 13.1 | 0.40 | 85 | 11.0 | 0.35 |
| 7 | 2 | 89 | 12.9 | 5.00 | 79 | 10.2 | 4.60 |
| 8 | 2 | 125 | 8.1 | 2.50 | 100 | 6.5 | 2.30 |
| 9 | 2 | 136 | 11.5 | 1.50 | 114 | 10.6 | 1.20 |
| 10 | 2 | 81 | 9.0 | 1.90 | 70 | 8.8 | 1.50 |
| 11 | 3 | 92 | 6.7 | 2.50 | 85 | 6.5 | 2.33 |
| 12 | 3 | 81 | 7.5 | 4.80 | 71 | 7.1 | 4.60 |
| 13 | 3 | 110 | 11.3 | 3.30 | 105 | 10.2 | 3.20 |
| 14 | 3 | 123 | 8.9 | 2.70 | 112 | 8.7 | 2.65 |
| 15 | 3 | 75 | 6.0 | 2.75 | 70 | 5.7 | 2.60 |

*(Data courtesy: Prof. K. Manjula, Department of Home Science, Sri Venkateswara University, Tirupati.)*

The analysis and discussion are however done for 30 records of Group-1.

**Analysis:**

This situation pertains to paired data where for each patient, the test is repeated at two time points and hence the data entry shall be in the same order in which it is recorded.

Let us create a data file in MS-Excel with a column heading bearing a tag 'PR' for pre-treatment values and 'PO' for post-treatment values. Let us define the level of significance as $\alpha = 0.05$. The null hypothesis is that the pre- and post-mean vectors do not differ in population (cohort).

We run the test with the following steps on an MS-Excel sheet.

*Step-1:* Select Real Statistics → Multivar → Hotelling T-Square

*Step-2:* For Array1 select the data for all the profile variables tagged as 'PO'

*Step-3:* For Array2 select the data for all the profile variables tagged as 'PR'

*Step-4:* Under the options select the radio button 'paired samples'

*Step-5:* Select a cell for displaying the output and press OK

This gives $\mathbf{T^2} = 4.7300$ with F = 1.4679 and p-value = 0.2454. Since p-value > 0.05 we accept the null hypothesis and conclude that there is no significant effect of the treatment on the thyroid panel.

It means the effect could also occur by chance. In this case there is no further need to find the CIs or check for component-wise differences. However, if the $\mathbf{T^2}$ test shows significance, the following procedure shall be executed.

Now let us define three new variables $d1 = (T3\_PO\text{-}T3\_PRE)$, $d2 = (T4\_PO\text{-}T4\_PRE)$ and $d3 = (TSH\_PO\text{-}TSH\_PRE)$ in the MS-Excel sheet which indicate for each case, the difference between the pre- and post-values. This constitutes a new panel of differences and the mean difference can be reported as a vector.

The mean and sample variance of these variables can be found with simple MS-Excel functions as shown in Figure 2.4.

The lower and upper confidence limits are obtained by using the variance of d1, d2 and d3 and the confidence coefficient given in Equation 2.6. The MS-Excel formula to do this is also shown in Figure 2.4.

| | | d1 | d2 | d3 |
|---|---|---|---|---|
| **Hotelling T-square Test** | | 1 | -0.2 | 0.0 |
| **Paired-sample test** | | -3 | 0.1 | -0.1 |
| | | -1 | -0.1 | 0.1 |
| T2 | 4.7300 | 2 | -0.4 | 0.1 |
| df1 | 3 | 2 | 0.5 | 0.1 |
| df2 | 27 | -1 | 0.5 | 0.1 |
| F | 1.4679 | -1 | 0 | 0.0 |
| p-value | 0.2454 | -1 | -0.1 | 0.0 |
| n | 30 | 0 | -0.1 | -0.1 |
| k | 3 | -1 | -0.1 | -0.2 |
| | | 3 | 0.1 | -0.1 |
| alpha | 0.05 | -2 | -0.1 | -0.2 |
| alpha-dash | 0.025 | -1 | -0.1 | -0.1 |
| **Mean diff** | | -0.2333 | -0.0467 | -0.0473 |
| **Variance** | | 7.7713 | 0.1102 | 0.0151 |
| **Conf.coef** | | 1.3555 | 0.1614 | 0.0597 |
| **Lower Conf Lt** | | -1.5888 | -0.2080 | -0.1070 |
| **Upper Conf Lt** | | 1.1221 | 0.1147 | 0.0124 |

FIGURE 2.4: MS-Excel worksheet for the paired sample Hotelling's $T^2$ test.

The mean differences and the confidence limits for the three components are shown below.

| Component | Mean Difference | 95% CIs | |
|---|---|---|---|
| | | Lower Limit | Upper Limit |
| T3 | -0.2333 | -1.5888 | 1.1221 |
| T4 | -0.0467 | -0.2080 | 0.1147 |
| TSH | -0.0473 | -0.1070 | 0.0124 |

According to the null hypothesis we set $\mathbf{H_0}$: $\boldsymbol{\mu_D} = \mathbf{\underline{0}}$, where $\boldsymbol{\mu_D} = \begin{bmatrix} \bar{d}_1 \\ \bar{d}_2 \\ \bar{d}_3 \end{bmatrix}$ and $\underline{\mathbf{0}} = \begin{bmatrix} 0 \\ 0 \\ 0 \end{bmatrix}$ is the null vector. By hypothesis, the mean difference for each component is zero and all three intervals shown above contain zero. Hence we accept that $\boldsymbol{\mu_D} = \mathbf{\underline{0}}$.

Consider another illustration.

**Illustration 2.4**  Consider the data from Group-2 of the dataset used in the Illustration 2.3.

For this data, repeating the same procedure as above we get $\mathbf{T}^2 = 97.3857$, $F = 30.2232$ and $p < 0.0001$. Hence there is a significant effect of treatment on the thyroid panel. The CIs are shown below.

| Component | Mean Difference | 95% CI Lower Limit | 95% CI Upper Limit | Significance |
|-----------|-----------------|--------------------|--------------------|--------------|
| T3        | -6.6000         | -9.3788            | -3.8212            | Yes          |
| T4        | -0.6800         | -0.9639            | -0.3961            | Yes          |
| TSH       | -0.2173         | -0.2996            | -0.1350            | Yes          |

Since each of the CIs does not contain zero, there is significant difference between the pre- and post-mean vectors.

We end this chapter with the observation that multivariate comparison of means is a valid procedure when the variables are correlated to each.

---

## Summary

Comparison of multivariate means is based on Hotelling's $\mathbf{T}^2$ test. We can compare the mean vector of a panel of variables with a hypothetical vector by using the one-sample $\mathbf{T}^2$ test whereas the mean vectors of two independent groups can be compared with the two-sample $\mathbf{T}^2$ test. The CI is another way of examining the truth of a hypothesis where we can accept the hypothesis, when the interval includes the hypothetical value. The *Real Statistics Add-ins* of MS-Excel has a module to perform various matrix operations as well as Hotelling's $\mathbf{T}^2$ test.

## Do it yourself (Exercises)

**2.1** The following are two matrices A and B.

$$A = \begin{bmatrix} 10.5 & 3.0 & 6.1 \\ 0 & 11.5 & 2.3 \\ 0 & 0 & 10.1 \end{bmatrix}, B = \begin{bmatrix} 3.2 & 0 & 0 \\ 3.6 & 4.1 & 0 \\ 2.1 & 4.3 & 6.2 \end{bmatrix}$$

Use MS-Excel functions to find a) The product AB of the matrices, b) A-inverse $(A^{-1})$, c) $A^{-1}B$ and d) $AB^{-1}$.

**2.2** With $k = 3$, $n = 10$, $\alpha = 0.05$, find the value of $F_{k,n-k,\alpha}$, using MS-Excel. (Hint: use FINV())

**2.3** Reconsider the data given in Table 2.1 with only 20 records. Take a panel of only two variables say BMI and HDL. Perform Hotelling's $\mathbf{T^2}$ test by taking a hypothetical mean vector $\begin{bmatrix} 28.0 \\ 35.0 \end{bmatrix}$.

**2.4** Compare the mean vectors of the following table with two groups A and B and draw conclusions.

| S.No | A | | | B | | |
|------|------|------|------|------|------|------|
| | x1 | x2 | x3 | x1 | x2 | x3 |
| 1 | 26.8 | 33 | 128 | 19.9 | 38 | 112 |
| 2 | 27.4 | 37 | 135 | 29.8 | 42 | 107 |
| 3 | 29.1 | 25 | 145 | 20.3 | 32 | 101 |
| 4 | 23.3 | 38 | 59 | 26.6 | 45 | 69 |
| 5 | 28.4 | 28 | 56 | 28.6 | 32 | 62 |
| 6 | 21.5 | 35 | 135 | 23.2 | 42 | 125 |
| 7 | 31.2 | 37 | 136 | 25.5 | 45 | 133 |
| 8 | 21.1 | 39 | 122 | 27.2 | 38 | 116 |
| 9 | 22.1 | 42 | 95 | 29.2 | 35 | 83 |
| 10 | 25.5 | 29 | 72 | 24.6 | 36 | 69 |
| 11 | 23.2 | 37 | 89 | 27.6 | 38 | 90 |
| 12 | 22.8 | 37 | 85 | 28.0 | 37 | 89 |
| 13 | 22.4 | 38 | 82 | 28.3 | 37 | 87 |
| 14 | 21.9 | 38 | 78 | 28.7 | 37 | 86 |
| 15 | 21.5 | 39 | 74 | 29.1 | 37 | 85 |
| 16 | 21.1 | 39 | 71 | 29.5 | 37 | 83 |
| 17 | 20.6 | 39 | 67 | 29.8 | 37 | 82 |
| 18 | 20.2 | 40 | 64 | 30.2 | 36 | 81 |
| 19 | 19.7 | 40 | 60 | | | |
| 20 | 19.3 | 41 | 57 | | | |

**2.5** Perform the paired comparison $\mathbf{T^2}$ test with the following data.

| S.No | Treated | | Control | |
|------|---------|---------|---------|---------|
|      | Pre     | Post    | Pre     | Post    |
| 1    | 20.2    | 22.8    | 18.3    | 21.4    |
| 2    | 19.7    | 22.4    | 17.9    | 21.9    |
| 3    | 19.3    | 21.9    | 18.4    | 19.5    |
| 4    | 18.9    | 21.5    | 18.0    | 21.1    |
| 5    | 18.4    | 21.1    | 17.6    | 20.6    |
| 6    | 18.0    | 20.6    | 17.1    | 20.2    |
| 7    | 17.6    | 20.2    | 16.7    | 19.7    |
| 8    | 17.1    | 19.7    | 16.3    | 19.3    |
| 9    | 16.7    | 19.3    | 15.8    | 18.9    |
| 10   | 16.3    | 18.9    | 15.4    | 18.4    |
| 11   | 15.8    | 18.4    | 14.9    | 18.0    |
| 12   | 15.4    | 18.0    | 14.5    | 17.6    |
| 13   | 14.9    | 17.6    | 14.1    | 17.1    |
| 14   | 14.5    | 17.1    | 13.6    | 16.7    |
| 15   | 14.1    | 16.7    | 13.2    | 16.3    |
| 16   | 13.6    | 16.3    | 12.8    | 15.8    |
| 17   | 13.2    | 15.8    | 12.3    | 15.4    |
| 18   | 12.8    | 15.4    |         |         |
| 19   | 12.3    | 14.9    |         |         |
| 20   | 11.9    | 14.5    |         |         |

## Suggested Reading

1. Alvin C.Rencher, William F. Christensen. 2012. *Methods of Multivariate Analysis.* 3rd ed. Brigham Young University: John Wiley & Sons.

2. Hummel, T.J., & Sligo, J.R. 1971. Empirical comparison of univariate and multivariate analysis of variance procedures. *Psychological Bulletin* 76(1): 49–57. DOI: 10.1037/h0031323.

3. Healy, M.J.R. 1969. Rao's paradox concerning multivariate tests of significance. *Biometrics* 25: 411–413.

4. Anderson T.W. 2003. *An Introduction to Multivariate Statistical Analysis* 3nd ed: Wiley Student Edition.

5. Box, G.E.P. 1949. Box's M-test. A general distribution theory for a class of likelihood criteria. *Biometrika* 36: 317–346.

6. Johnson, R.A., & Wichern, D.W. 2014. *Applied multivariate statistical analysis,* 6th ed. Pearson New International Edition.

7. Sarma K.V.S. 2010. *Statistics Made Simple-Do it yourself on PC.* 2nd edition. Prentice Hall India.

# Chapter 3

# Analysis of Variance with Multiple Factors

3.1   Review of Univariate Analysis of Variance (ANOVA) ...........   51
3.2   Multifactor ANOVA  .............................................   55
3.3   ANOVA with a General Linear Model ..........................   57
3.4   Continuous Covariates and Adjustment  .......................   61
3.5   Non-Parametric Approach to ANOVA  ..........................   65
3.6   Influence of Random Effects on ANOVA  .......................   68
      Summary ......................................................   69
      Do it yourself (Exercises)  ....................................   69
      Suggested Reading  ............................................   71

> The analysis of variance is not a mathematical theorem, but rather a convenient method of arranging the arithmetic.
>
> Sir Ronald A. Fisher (1890 – 1962)

## 3.1   Review of Univariate Analysis of Variance (ANOVA)

The Analysis of Variance (ANOVA) is a procedure to compare the mean values of a single continuous variable across several groups (populations) based on the data obtained from samples drawn from these groups. The response variable Y is assumed to be influenced by several *factors* like dose of a treatment, age group, social status etc., each of which is *categorical* with pre-defined fixed levels. The total variation in the data on Y can be attributed to the variation due to each factor and its significance will be studied by using ANOVA.

If there is only one factor affecting Y, we use *One-way ANOVA* in which the *main effect* of the factor is compared. When two or more independent factors like age group, gender, disease-stage etc., are present, we have to carry out a *Multifactor ANOVA* where not only the main effect of each factor is tested for significance, but the *interaction* between factors (called *combined effect*) is also studied. We review these methods with a focus on computation skills and interpretation of results.

**One-way ANOVA:**

Let there be a single factor like *age group* having k levels and we wish to check whether the mean response remains the same in all the k groups. Since the data is classified according to only one factor, it is called *one-way classified* data and we use the term *One-way ANOVA* which is based on the following assumptions:

1. The data on Y in the $i^{th}$ group is normally distributed.

2. The variance of Y in each of the k groups is the same.

Let the mean of Y in the $k^{th}$ group be denoted by $\mu_k$. Then we wish to test the null hypothesis $H_0$: $\mu_1 = \mu_2 = \ldots = \mu_k$ against the alternative hypothesis $H_1$: *At least two means (out of k) are not equal.* The second assumption $\sigma_1^2 = \sigma_2^2 = \ldots = \sigma_k^2 = \sigma^2$ is known as *homoscedasticity*.

The truth of $H_0$ is tested by comparing the ratio of a) Mean Sum of Squares (MSS) due to the factor to b) Residual MSS. This variance-ratio follows Snedecor's F-distribution and hence the hypothesis is tested by using an F-test. If the p-value of the test is less than the notified error rate $\alpha$, we reject the null hypothesis and conclude that there is a significant effect of the factor on the response variable Y.

If the null hypothesis is rejected, the next job is to find out which of the k-levels is responsible for the difference. This is done by performing a *pairwise comparison test* of group means such as Duncan's Multiple Range Test (DMRT), Tukey's test, Least Significant Difference (LSD) test or Scheffe's test. All these tests look very similar to the Student's t-test but they make use of the residual MSS obtained in the ANOVA and hence such tests are done only after performing the ANOVA. The Dunnett's test is used when one of the levels is taken as *control* and all other means are to be compared with the mean of the control group.

The calculations of ANOVA can be performed easily with statistical software. In general, data is created as a flat file with all groups listed one below the other. However, in order to perform ANOVA using the Data Analysis Pak of Excel or Real Statistics Resource pack, the data has to be entered in separate groups (columns) each for one level of the factor. This however requires additional effort to *redesign* the data, which demands significant time and effort.

Alternatively, we can use SPSS with more options convenient for input and output. The following steps can be used for analysis.

*Step-1:* Open the SPSS data file

*Step-2:* Choose Analyze → Compare Means → One way ANOVA

*Step-3:* Select the outcome variable into the 'dependent variable' box

*Step-4:* Select the name of the factor into the 'Factor' box

*Step-5:* Click on the *options* tab and select *descriptive statistics*

*Step-6:* Click on the *Post Hoc* tab and select one of the required tests (like DMRT)

*Step-7:* Check the *means plot*. Press OK

This produces the output broadly having a) Mean and S.D of the each group and b) the ANOVA table.

**Remark-1:**

a) While choosing factors, we shall make sure that the data on the factor has a few discrete values and not a stream of continuous values. For instance if actual *age* of the patients is recoded into 4 groups then *age group* is a meaningful factor but not actual *age*.

b) When the F-test in the ANOVA shows no significant effect of the factor, then Post Hoc tests are not required. We may switch over to Post Hoc only when ANOVA shows significance.

c) Suppose there are 4 levels of a factor, say placebo, 10mg, 15mg and 20mg of a drug. If we are interested in comparing each response with that of the control, we shall use Dunnett's test and select the control group as 'first' to indicate placebo; or 'last' if the control group is at the end of the list.

Sarma (2010) discussed methods of running ANOVA with MS-Excel and also SPSS.

Consider the following illustration.

**Illustration 3.1** In an endocrinology study a researcher has measured the Bone Mineral Density (BMD) of each patient at three positions viz., *spine*, *neck of femur* and *greater trochanter of femur* along with some other parameters like age, gender (1 = Male, 2 = Female), Hypertension (HTN), 0 = No, 1 = Yes) and Body Mass Index (BMI).

A new variable *age group* was created with three groups (i) ⩽30 yrs, (ii) 31–40 yrs and (iii) 41 & above, coded as 1, 2, 3 respectively. For ease in handling, we denote the panel variables as *bmd_s*, *bmd_nf* and *bmd_gtf* respectively.

A portion of the data with 15 records is shown in Table 3.1 but analysis is based on complete data with 40 records. We call this data *BMD data* for further reference.

TABLE 3.1: Bone Mineral Density of 15 patients

| S.No | Age | Gender | BMI | Age group | HTN | bmd_s | bmd_nf | bmd_gtf |
|------|-----|--------|-----|-----------|-----|-------|--------|---------|
| 1 | 34 | 1 | 24 | 2 | 1 | 0.933 | 0.820 | 0.645 |
| 2 | 33 | 2 | 22 | 2 | 1 | 0.889 | 0.703 | 0.591 |
| 3 | 39 | 2 | 29 | 2 | 0 | 0.937 | 0.819 | 0.695 |
| 4 | 32 | 2 | 20 | 2 | 1 | 0.874 | 0.733 | 0.587 |
| 5 | 38 | 2 | 25 | 2 | 1 | 0.953 | 0.824 | 0.688 |
| 6 | 37 | 2 | 13 | 2 | 0 | 0.671 | 0.591 | 0.434 |
| 7 | 41 | 2 | 23 | 3 | 1 | 0.914 | 0.714 | 0.609 |
| 8 | 24 | 2 | 23 | 1 | 1 | 0.883 | 0.839 | 0.646 |
| 9 | 40 | 2 | 24 | 2 | 1 | 0.749 | 0.667 | 0.591 |
| 10 | 16 | 1 | 18 | 1 | 1 | 0.875 | 0.887 | 0.795 |
| 11 | 43 | 1 | 21 | 3 | 1 | 0.715 | 0.580 | 0.433 |
| 12 | 41 | 2 | 18 | 3 | 1 | 0.932 | 0.823 | 0.636 |
| 13 | 39 | 1 | 16 | 2 | 0 | 0.800 | 0.614 | 0.518 |
| 14 | 45 | 2 | 17 | 3 | 1 | 0.699 | 0.664 | 0.541 |
| 15 | 42 | 2 | 25 | 3 | 0 | 0.677 | 0.547 | 0.497 |

*(Data courtesy: Dr. Alok Sachan, Department of Endocrinology, Sri Venkateswara Institute of Medical Sciences (SVIMS), Tirupati.)*

Now it is desired to check whether the average BMD is the same in all the age groups.

**Analysis:**

Using *age group* as the single factor, we can run one-way ANOVA from SPSS which produces the results shown in Table 3.2.

By default $\alpha$ is taken as 0.05. We note that ANOVA will be run 3 times, one time for each variable. The F values and p-values are extracted from the ANOVA table. SPSS reports the p-value as 'sig'.

It is possible to study the effect of *age group* on each panel variable separately by using univariate ANOVA. We will first do this and later in Chapter 4 we see that a technique called *multivariate ANOVA* is more appropriate. What is done in this section is only univariate one-way analysis.

Since the p-value is higher than 0.05 for all three variables, we infer that there is no significant change in the mean values of these variables. For this reason we need not carry out a multiple comparison test among the means at the three age groups.

TABLE 3.2: One-way ANOVA

| Variable | Age group | N | Mean | Std. Dev | F | p-value |
|---|---|---|---|---|---|---|
| bmd_s | ⩽30 yrs | 14 | 0.884 | 0.054 | 0.674 | 0.516 |
| | 31-40 yrs | 14 | 0.840 | 0.088 | | |
| | 41 & above | 12 | 0.829 | 0.208 | | |
| bmd_nf | ⩽30 yrs | 14 | 0.751 | 0.075 | 2.002 | 0.149 |
| | 31-40 yrs | 14 | 0.695 | 0.086 | | |
| | 41 & above | 12 | 0.672 | 0.145 | | |
| bmd_gtf | ⩽30 yrs | 14 | 0.633 | 0.079 | 1.449 | 0.248 |
| | 31-40 yrs | 14 | 0.569 | 0.082 | | |
| | 41 & above | 12 | 0.586 | 0.144 | | |

**Remark-2:**

Instead of *age group*, the effect of *gender* and HTN on the outcome variables is to be examined and one has to perform an independent sample t-test for each variable separately between the two groups, since each of these two factors have only two levels and ANOVA is not required.

Now let us consider the situation where there are two or more factors that affect the outcome variable Y. This is done by running a multifactor ANOVA.

## 3.2 Multifactor ANOVA

Suppose the response Y is affected by more than one factor. For instance let age group, HTN and Physical Activity be three factors that are thought to influence Y and these are denoted by A, B and C respectively. Then there will be three main effects, A, B and C, whose significance can be tested.

In addition, there will be 'joint combined effects' called interactions like AB, BC, AC and ABC. We can now test one hypothesis on each main effect/interaction to assess their statistical significance.

The statistical approach to handle this problem is different from the one-way ANOVA approach.

The procedure is based the concept of a *general linear model* in which the data on the dependent variable (Y) from each individual can be explained by an additive model

$$Y = \text{Constant} + \alpha_i + \beta_{ij} + \text{random error}$$

where $\alpha_i$ = effect of the $i^{th}$ factor and $\beta_{ij}$ = interaction of the $i^{th}$ and $j^{th}$ factors. The term *constant* represents the baseline value of Y irrespective of the effect of these factors.

The analysis is based on the principle of a *linear model* where each factor or interaction will have an additive effect on Y. Each effect is estimated in terms of the mean values and tested for significance by comparing the variance due to the factor/interaction with that of the residual/unexplained variation, with the help of an F-test.

The problem is to estimate these effects from the sample data and test the significance of the observed effects. The general linear model is a unified approach, combining the principle of ANOVA and that of *Linear Regression* (discussed in Chapter 4). As a part of ANOVA, a linear model will be fitted to the data and tested for its significance.

A brief outline of the important characteristics of the fitted model is given below as a caution before proceeding further in multifactor ANOVA.

a) The adequacy (goodness) of the regression model is expressed in terms of a measure $R^2$, which takes values between 0 and 1.

b) A better measure of adequacy of the model is the adjusted $R^2$ which can be derived from $R^2$ (more details in Chapter 4).

c) In both cases, a higher value of $R^2$ indicates better fit of the model to the data.

d) For instance $R^2 = 0.86$ indicates that 86% of the behavior of Y can be explained by the model. Again $R^2 = 0.09$ refers to a poor model questioning the choice of linear model or the choice of factors in the model.

At the end of analysis we can answer the following questions.

1. Whether a particular factor shows a significant effect on Y?

2. Whether there is any interaction of factors showing a significant effect on Y?

3. If any interaction is significant, which combination of factor levels would yield the best outcome?

4. What are the estimated mean values of Y at different levels of a factor after adjusting for the covariates, if any?

In the following section we elaborate on ANOVA using the general linear model.

## 3.3   ANOVA with a General Linear Model

The multifactor ANOVA requires many computations and computer software helps to do it. We explain the approach with SPSS. After understanding the inputs required for using the SPSS options we focus on the type of output and its interpretation with the help of an illustration.

The SPSS menu sequence is *Analyze → General Linear Model → Univariate*

The input options include the following:

1. **Dependent Variable:** Y(Response/Outcome).

2. **Fixed Factors:** All the factors which would influence Y and having known fixed levels.

3. **Random Factors:** All the factors whose values (levels) are considered as a random selection from a larger set of categories.

4. **Covariates:** Continuous variables which would have a partial influence on Y. We get the 'adjusted mean' of Y by eliminating the marginal influence of each covariate on Y. For instance, *age-adjusted bmd_nf* is one such result.

5. **Model**: We may select the option *full factorial* which means that all possible effects including interactions will be considered in the model. Alternatively, we may fix only a few effects under the option *custom*. Under the label *interaction* we can select joint effects of two or more factors.

6. **Intercept:** When there is a non-zero base line value, we need intercept to be included in the model; otherwise we can set intercept as zero. As an option we may avoid having intercept in the model. But it is suggested to keep the option checked.

7. **Options:** We can choose to display descriptive statistics for factors and interactions. Mean values can be displayed and compared across levels of the factor(s). We can also choose a test for homogeneity of variances

to know whether the variances are equal within each factor. This is done by Levene's test for which p > 0.05 indicates homogeneity of variances.

Consider the following illustration.

**Illustration 3.2**  We will reconsider the BMD data used in Illustration 3.1 and study the effect of *gender* and HTN on the outcome variables. Since we can study only one variable at a time in univariate analysis with multiple factors, we consider *bmd_nf* with the following model.

$$bmd\_nf = \text{constant} + \text{effect of } gender + \text{effect of HTN} +$$
$$\text{effect of } (gender \text{ \& HTN}) + \text{random error} \tag{3.1}$$

We wish to estimate the effect of HTN, *gender* and their interaction on *bmd_nf* and test for significance.

**Analysis:**

In order to run the ANOVA we choose the following options as shown in Figure 3.1.

a) Dependent variable → *bmd_nf*

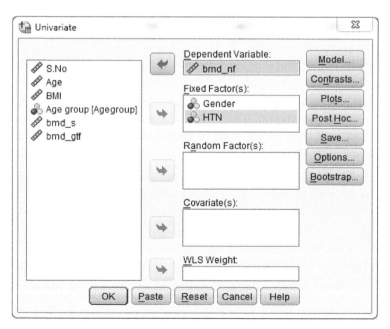

FIGURE 3.1: GLM–univariate-selection of variables.

b) Fixed factors *gender*, HTN

c) Model → Full factorial

d) Include intercept in the model = Yes

e) Options → Estimated Marginal Means, Display means for → *gender* * HTN, Homogeneity tests.

The outcome of the ANOVA is shown in Table 3.3

TABLE 3.3: Multifactor ANOVA (Univariate)

Tests of Between-Subjects Effects
Dependent Variable: bmd_nf

| Source | Type III Sum of squares | df | Mean Square | F | p-value |
|---|---|---|---|---|---|
| Corrected Model | 0.102[a] | 3 | 0.034 | 3.589 | 0.023 |
| Intercept | 13.037 | 1 | 13.037 | 1371.678 | 0.000 |
| Gender | 0.002 | 1 | 0.002 | 0.199 | 0.658 |
| HTN | 0.062 | 1 | 0.062 | 6.546 | 0.015 |
| Gender * HTN | 0.052 | 1 | 0.052 | 5.492 | 0.025 |
| Error | 0.342 | 36 | 0.010 | | |
| Total | 20.482 | 40 | | | |
| Corrected Total | 0.444 | 39 | | | |

[a]$R$ Squared = 0.230 (Adjusted R Squared = 0.166).

The output of ANOVA (detailed tables not shown here) is understood as follows.

a) ANOVA is basically a test for comparing 'factor means', when the number of factors is more than two. Due to the mathematical structure of the problem, this test reduces to a variance-ratio test (F-test). The variance due to known factors is compared with the unexplained variation.

b) The term *corrected model* indicates a test for the goodness of fit of the assumed linear model in the ANOVA. In this case, the model has $F = 3.589$ with $p = 0.023$ ($< 0.05$) which means that the model is statistically significant. However $R^2 = 0.230$ means only 23.0% of the behavior of the *bmd_nf* is explained in terms of the selected factors and their interactions. This could be due to several reasons. For instance, there might be a non-linear relationship of Y with the factors but we have used only a linear model. There could also be some other influencing factors, not considered in this model.

c) We have chosen to keep intercept in the model which reflects the fact that there will be some baseline value of *bmd_nf* in every individual, irrespective of the status of HTN and *gender*. For this reason we prefer to keep the constant in the model and its presence is also found to be significant.

***Caution***: If we remove the intercept from the model we get $R^2 = 0.983$ with F = 529.761 (p < 0.0001) which signals an excellent goodness of fit. Since the model in this problem needs a constant, we should not remove it. Hence comparison of this $R^2$ value with that obtained earlier is not correct.

d) The Levene's test shows p = 0.948 which means that the null hypothesis of equal variances across the groups can be accepted.

e) The F-value for each factor is shown in the table along with the p-value. The main effect of HTN is significant, while that of *gender* is not significant.

f) Interestingly, the interaction (joint effect) of HTN & *gender* is found to be significant with p = 0.025. This shows that *gender* individually has no influence on *bmd_nf* but shows an effect in the presence of HTN.

g) Since the interaction is significant, we have to present the descriptive statistics to understand the mean values as envisaged by the model. We can use the option *descriptive statistics* which shows the mean and standard deviation of *bmd_nf* given for each group. A better method is to use the option *estimated marginal means* which gives mean along with the SE and CIs. In both cases the means remain the same but the second method is more apt because we are actually estimating the mean values through the linear model and hence these estimates shall be displayed along with SE. The results are shown in Table 3.4.

TABLE 3.4: Estimated marginal means of *bmd_nf*

| HTN | Male | | Female | |
|-----|------|-----------|------|-----------|
| | Mean | Std. Error | Mean | Std. Error |
| No | 0.615 | 0.056 | 0.687 | 0.028 |
| Yes | 0.801 | 0.034 | 0.695 | 0.024 |

It follows from the above table that those males without hypertension have a higher *bmd_nf*.

h) Multifactor analysis ends with estimation of effect sizes mentioned in the model Equation 3.1. This is done by selecting the option 'parameter estimate'. In this case the estimated model becomes

$$bmd\_nf = 0.695 + 0.105 * gender - 0.008 * \text{HTN}$$
$$+ 0.177 * (gender \ \& \ \text{HTN})$$

These values represent the constant (baseline value) and the marginal effect of *gender*, HTN and *gender* * HTN respectively and given in column 'B' of the table of 'parameter estimates'. When the interaction term is not consider, the marginal effect will be the difference of the estimated marginal means due to a factor like *gender* or HTN.

**Remark-3:**

Though the validity of assumptions of ANOVA has to be checked before analysis, small deviations from the assumptions do not drastically change the results.

In the following section we study the influence of continuous variables (as covariates) on response Y.

---

## 3.4 Continuous Covariates and Adjustment

Sometimes there will be factors measured on a continuous scale like Age, BMI, HbA1c etc., which are called *covariates* and their influence on the outcome variable could be of interest. They may have significant correlation with the outcome Y. Since they are continuous (not categorical) they will not fit into the ANOVA model like *categorical factors*. Analysis with such cofactors is called *Analysis of Covariance* (ANCOVA).

Now the effect of already included factors like HTN will also carry the marginal influence of the covariate which inflates/deflates the true effect. For this reason we can include one or more covariates into the model and estimate the main effects of factors after *adjusting for the covariates*. Adjustment means removal of the marginal effect of the covariate(s) on the outcome Y and then estimating the real effect.

ANCOVA is performed by using a general linear model as done in Section 3.3 except that covariates are selected into the model along with fixed factors.

Consider the following illustration.

**Illustration 3.3** Reconsider the data used in Illustration 1.1. A subset of the data is shown in Table 3.5 with variables *age*, *gender* and CIMT.

We wish to examine whether RA status and *gender* or their interaction have any significant effect on CIMT. We have relabeled the variable Group as RA ( 1 = "Yes" and 2 = "No"). The analysis and discussion is however based on complete data.

TABLE 3.5: Sample data on CIMT

| S.No | AS | Age | Gender | CIMT | S.No | AS | Age | Gender | CIMT |
|------|----|-----|--------|------|------|----|-----|--------|------|
| 1 | 1 | 56 | 1 | 0.600 | 11 | 0 | 45 | 0 | 0.460 |
| 2 | 1 | 43 | 0 | 0.570 | 12 | 0 | 37 | 0 | 0.440 |
| 3 | 1 | 45 | 0 | 0.615 | 13 | 0 | 63 | 0 | 0.530 |
| 4 | 1 | 39 | 0 | 0.630 | 14 | 0 | 41 | 0 | 0.510 |
| 5 | 1 | 31 | 0 | 0.470 | 15 | 0 | 51 | 0 | 0.410 |
| 6 | 1 | 34 | 0 | 0.630 | 16 | 0 | 40 | 0 | 0.560 |
| 7 | 1 | 37 | 1 | 0.515 | 17 | 0 | 42 | 0 | 0.480 |
| 8 | 1 | 54 | 0 | 0.585 | 18 | 0 | 45 | 1 | 0.510 |
| 9 | 1 | 59 | 0 | 0.615 | 19 | 0 | 38 | 0 | 0.510 |
| 10 | 1 | 33 | 0 | 0.415 | 20 | 0 | 43 | 1 | 0.680 |

**Analysis:**

Applying the general linear model as done in the previous illustration, we find that the model with RA and *gender* has a significant effect on CIMT but the interaction (joint effect) is not significant. The model is statistically significant with $R^2 = 0.265$, $F = 7.202$ and $p < 0.0001$. The intercept plays a significant role in the model indicating that the estimate of the overall CIMT is 0.583 with 95% CI [0.548, 0.619] irrespective of the effect of the status of RA and *gender*.

At this point it is worth understanding the estimated values of CIMT as mean ± SE as follows.

a) Gender effect:

   (i) Female: $0.534 \pm 0.014$ and (ii) Male: $0.632 \pm 0.033$

b) RA status effect:

   (i) RA-Yes:$0.635 \pm 0.025$ and (ii) RA-No: $0.532 \pm 0.025$

c) Gender & RA interaction is given below:

| Gender | RA | |
|--------|-----|-----|
|  | Yes | No |
| Female | $0.581 \pm 0.020$ | $0.487 \pm 0.020$ |
| Male | $0.689 \pm 0.046$ | $0.576 \pm 0.046$ |

We observe the following.

1. Irrespective of the RA status, males have higher CIMT.

2. Irrespective of *gender*, patients with RA have higher CIMT.

3. Though the interaction is not significant, we may observe that Male patients having CIMT are expected to have the highest CIMT = 0.689.

Suppose we wish to understand the influence of the actual age of the patient on CIMT.

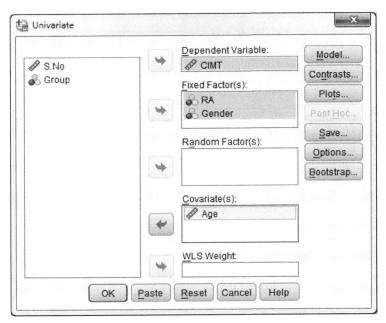

FIGURE 3.2: GLM–univariate-selection of covariate.

This can be can be included in the model by sending the variable *age* into the covariate box as shown in Figure 3.2.

TABLE 3.6: Tests of Between-Subjects Effects with covariates

Dependent Variable: CIMT

| Source | Type III Sum of squares | df | Mean Square | F | p-value |
|---|---|---|---|---|---|
| Corrected Model | 0.402[a] | 4 | 0.100 | 12.36 | <0.001 |
| Intercept | 0.264 | 1 | 0.264 | 32.526 | <0.001 |
| Age | 0.168 | 1 | 0.168 | 20.729 | <0.001 |
| RA | 0.063 | 1 | 0.063 | 7.775 | 0.007 |
| Gender | 0.067 | 1 | 0.067 | 8.253 | 0.006 |
| RA * Gender | 0.000 | 1 | 0.000 | 0.014 | 0.905 |
| Error | 0.479 | 59 | 0.008 | | |
| Total | 20.206 | 64 | | | |
| Corrected Total | 0.881 | 63 | | | |

[a]R Squared = 0.456 (Adjusted R Squared = 0.419).

With this selection, the ANOVA model with *age*, *gender* and RA status is found to be significant. The model has $R^2 = 0.456$, F = 12.360 and p < 0.001.

This is a better model than the earlier one ($R^2 = 0.265$) where *age* was not taken into account. The ANOVA is shown in Table 3.6.

The estimated marginal means adjusted for *age* are shown in Table 3.7.

TABLE 3.7: Table of means for *RA* & gender interaction adjusted for age

Dependent Variable: CIMT

| RA | Gender | Mean | Std.Error | 95% CI | |
|----|--------|------|-----------|--------|--|
| | | | | Lower Bound | Upper Bound |
| Yes | Female | 0.581[a] | 0.017 | 0.546 | 0.616 |
| | Male | 0.667[a] | 0.041 | 0.585 | 0.748 |
| No | Female | 0.490[a] | 0.017 | 0.455 | 0.525 |
| | Male | 0.583[a] | 0.040 | 0.502 | 0.664 |

[a]Covariates appearing in the model are evaluated at the following values: *age* = 47.36.

It is easy to note that the adjusted means for the four combinations in the above table are different from those obtained without including the covariate.

We now present the means ± SE of CIMT with respect to the main effects and interactions.

a)  RA status effect (adjusted *age*) i) RA-Yes: $0.624 \pm 0.022$ and ii) RA-No: $0.537 \pm 0.022$.

b)  Gender effect (adjusted *age*) i) Female: $0.536 \pm 0.012$ and ii) Male: $0.625 \pm 0.029$.

c)  Gender & RA status effect (adjusted *age*).

| Gender | RA | |
|--------|-----|--|
| | Yes | No |
| Female | $0.581 \pm 0.017$ | $0.490 \pm 0.017$ |
| Male | $0.667 \pm 0.041$ | $0.583 \pm 0.040$ |

**Remark-4:**

Comparing the estimated means and their SEs with and without a covariate we observe that the SE of the mean is in general lower when a covariate is used, than without a covariate.

In conclusion, we notice that by using the general linear model, we can not only test the effect of categorical factors but also include covariates and obtain reliable estimates of the effects adjusted for the covariates.

In the following section we discuss a non-parametric approach to the ANOVA which is useful when the assumptions of ANOVA are not valid for a dataset.

## 3.5 Non-Parametric Approach to ANOVA

Sometimes the data on the outcome variable (Y) will not be normally distributed in different categories of the factor (like *gender*, *age group*, treatment arm etc.) under consideration. It is also possible that the standard deviations in different categories will not be the same (widely different). This violates the assumption of normality of residuals and also failure of the assumption of homoscedasticity (equality of within group variances). In some other instances, the data will be available on an ordinal scale or the data is already ranked.

By using suitable transformations, normality can be induced into the model and in some cases variances can be stabilized so that the regular ANOVA and the general linear model can be used.

As an alternative we can use a non-parametric method of comparing the 'mean rank' of Y among the groups with the help of a test called the *Kruskal Wallis* test. Non-parametric tests are also known as *distribution-free methods* because no assumptions are made, for these tests, on the population from which the samples are drawn. The Kruskal Wallis test is considered as the non-parametric equivalent of the one-way ANOVA and the procedure is based on the ranks assigned to the Y values.

SPSS contains a group of tools for non-parametric methods and the Krushkal Wallis test can be chosen with the options: Analyze → Non-parametric Tests → Independent Samples. Selecting the customize options for the tests, we can input the Y variable and the factor (only one factor!) into the model. The settings menu shows the Kruskal Wallis test. As an option we can choose descriptive statistics to know the mean and S.D of each variable. The output contains a table showing the test results and the null hypothesis. We can also select more than one outcome variable and the test results appear separately.

Consider the following illustration.

**Illustration 3.4** Serum Homocysteine is an important amino acid in the blood, often used as a factor for heart disease. The data shown in Table 3.8 contains the Serum Homocysteine (S_Hcy) and Serum Creatinne(S_Cr) levels of 30 anemic patients classified as a) Iron Deficiency Anemia (Category = 1), b) B12 Deficiency Anemia (Category = 3) and c) Both (Category = 2).

TABLE 3.8: Homocysteine data among patients suffering from anemia

| S.No | S_Hcy | S_Cr | Category | S.No | S_Hcy | S_Cr | Category |
|------|-------|------|----------|------|-------|------|----------|
| 1 | 16 | 0.80 | 1 | 16 | 21 | 0.70 | 2 |
| 2 | 16 | 0.46 | 1 | 17 | 21 | 0.69 | 2 |
| 3 | 15 | 0.83 | 1 | 18 | 20 | 0.52 | 2 |
| 4 | 15 | 1.60 | 1 | 19 | 19 | 0.86 | 2 |
| 5 | 14 | 0.72 | 1 | 20 | 17 | 0.75 | 3 |
| 6 | 13 | 0.82 | 1 | 21 | 18 | 1.95 | 3 |
| 7 | 13 | 0.75 | 1 | 22 | 24 | 0.62 | 3 |
| 8 | 13 | 0.71 | 1 | 23 | 20 | 0.45 | 3 |
| 9 | 13 | 1.01 | 1 | 24 | 20 | 0.71 | 3 |
| 10 | 13 | 0.90 | 1 | 25 | 42 | 0.32 | 3 |
| 11 | 32 | 0.70 | 2 | 26 | 17 | 0.70 | 3 |
| 12 | 28 | 0.34 | 2 | 27 | 17 | 0.39 | 3 |
| 13 | 24 | 0.90 | 2 | 28 | 16 | 0.39 | 3 |
| 14 | 23 | 1.30 | 2 | 29 | 20 | 0.34 | 3 |
| 15 | 21 | 0.70 | 2 | 30 | 21 | 0.65 | 3 |

*(Data courtesy: Dr. C. Chandra Sekhar, Department of Haemotology (SVIMS), Tirupati.)*

We wish to test whether the Homocysteine and Creatinine levels differ significantly among the three anemic groups.

**Analysis:**

The data can be taken into an SPSS file in the same format as shown in Table 3.5. The SPSS options are Analyse → Non-parametric tests → Independent Samples → Customize analysis. Then select option Fields and choose the variable Serum_Hcy and group as category. Then select the options 'settings' and choose 'customize tests'. Then select the option 'Kruskal_Wallis 1-Way ANOVA (k samples)'.

Running these commands produces the output as shown in Table 3.9.

TABLE 3.9: Output of Kruskal Wallis one-way ANOVA

| Null Hypothesis | Test | Sig. | Decision |
|-----------------|------|------|----------|
| The distribution of Serum_Homocysteine is the same across categories of Category. | Independent-Samples Kruskal-Wallis Test | 0.000 | Reject the null hypothesis. |

Asymptotic significances are displayed. The significance level is 0.05.

Unlike the classical ANOVA, this test is based on ranking of Y values in different groups. The null hypothesis is that the Homocysteine values have the same pattern in all the three groups under consideration. We shall note that there is no reference to the mean or median for comparison and further there is no assumption about the underlying distribution of data. The decision says that the hypothesis rejected which means that the Homocysteine value can be considered different in different groups.

As a post hoc analysis we wish to know which pair of groups has contributed to the rejection of the null hypothesis. This is done by a double click on the output table, which produces a multiple comparison test with graphic visualisation as shown in Figure 3.3.

## Pairwise Comparisons of Category

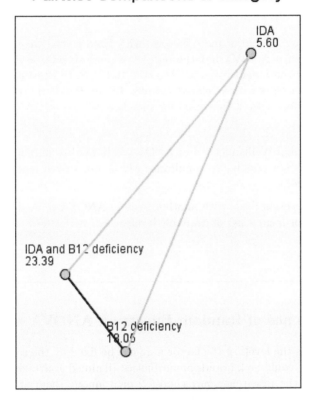

Each node shows the sample average rank of Category.

FIGURE 3.3: Graphic display of multiple comparisons.

Further, Table 3.10 shows the results of the paired comparison test.

TABLE 3.10: Non-parametric paired comparison test

| Sample1-Sample2 | Test Statistic | Std. Error | Std. Test Statistic | Sig. | Adj.Sig. |
|---|---|---|---|---|---|
| IDA-B12 deficiency | -12.445 | 3.825 | -3.254 | 0.001 | 0.030 |
| IDA-IDA and B12 deficiency | -17.789 | 4.022 | -4.423 | 0.000 | 0.000 |
| B12 deficiency-IDA and B12 deficiency | 5.343 | 3.935 | 1.358 | 0.174 | 0.523 |

Each row tests the null hypothesis that the Sample1 and Sample2 distributions are the same.
Asymptotic significances are displayed. The significance level is 0.05.

When there are two or more factors (with fixed levels) we can use a test called Friedman's ANOVA but the output of analysis will be different from that of the conventional two-factor ANOVA. In SPSS, this test is found under non-parametric tests with k-related samples. In fact the data arrangement for this test is different from that used for the classical two-way ANOVA.

**Remark-5:**

The Krushkal Wallis method of ANOVA is limited to only a single factor at a time. It is not possible to simultaneously handle two or more factors and their interactions.

The next section deals with another issue in ANOVA related to a random choice of factor from a list of available levels.

## 3.6    Influence of Random Effects on ANOVA

Sometimes the levels of the factor may not be fixed by the researcher. For instance there could be k brands of antibiotics all aimed at attacking a specific infection but the researcher may choose 2 or 3 among them at random and conduct a study. The average effect of the antibiotic on the response will be compared among the three groups but there are other brands of drugs not covered by the study. In such cases the factor levels selected by the researcher represent only a random sample of the exhaustive list of levels.

A random effects model is a statistical assessment of the mean response at the selected levels of the factor but not all levels. As such, the generalization of factor effect requires knowledge about the SE due to random selection of levels.

The ANOVA with a single or multiple random effects produces for each effect, a quantity called *Expected Mean Squares (EMS)* which estimates the combined variance due to the effect and that of the residual. The calculations are similar to that of a fixed effects model. However, when the F-value shows significance for an effect, we wish to estimate how much of the total variation in the response can be attributed to the factor under consideration and how much can be left to random (unknown) causes. Many statistical software packages contain a module to handle random effect in ANOVA as well as linear regression models.

With this we end the discussion on the univariate ANOVA with multiple categorical factors as well as continuous covariates.

---

## Summary

ANOVA is a powerful tool for comparing the mean values of a variable across several groups. We can estimate the effect of many factors (having discrete levels) and their interactions on the outcome of the experiment and test for their statistical significance. Sometimes the response is partly influenced by one or more continuous variables also and we have to take into account the effect of such factors called *covariates*. This would improve the quality of the estimates of the effects and also provide more meaningful decisions on the significance. Multiple-comparison of group means is a post hoc procedure, initiated when the factors (at 3 or more levels) show significance. While we have discussed all these aspects, we have observed that a non-parametric approach also exists to perform a test like ANOVA.

---

## Do it yourself (Exercises)

**3.1** The following data refers to the measurements recorded on a chemical process (in suitable units) by three different methods (A,B,C) by using 4 types of detergents. For each combination of detergent and method, two observations are made.

It is desired to test whether the two factors viz., detergent and method have any effect on the response. Further is there is any significant interaction of method and detergent on the response?

| Detergent | Method | | |
|:---:|:---:|:---:|:---:|
| | A | B | C |
| 1 | 45,46 | 43,44 | 51,49 |
| 2 | 47,48 | 46,48 | 52,50 |
| 3 | 48,49 | 50,52 | 55,52 |
| 4 | 42,41 | 37,39 | 49,51 |

(Hint: Create an Excel file with three columns, Det, Met and Response. Recode A, B, C into 1, 2, 3. The combination Det = 1 and Met = 1 appears in two times with response 5 and 46 respectively. Create this file in SPSS and run two-factor ANOVA.)

**3.2** The following data refers to the BMI (Kg/m$^2$) of 24 individuals under treatment for a food supplement to reduce BMI. The two interventions are a) Physical Exercise (code = 1) and b) Physical Exercise plus food supplement (code = 2). The other factors are Gender (1 = male, 2 = female) and Age (years).

| S.No | Age | Gender | Intervention | BMI | S.No | Age | Gender | Intervention | BMI |
|:---:|:---:|:---:|:---:|:---:|:---:|:---:|:---:|:---:|:---:|
| 1 | 28 | 1 | 1 | 25.87 | 13 | 50 | 1 | 2 | 24.28 |
| 2 | 45 | 2 | 1 | 25.68 | 14 | 33 | 1 | 2 | 21.32 |
| 3 | 54 | 1 | 1 | 27.73 | 15 | 40 | 2 | 2 | 21.47 |
| 4 | 55 | 2 | 1 | 28.36 | 16 | 28 | 2 | 2 | 20.96 |
| 5 | 55 | 1 | 1 | 28.99 | 17 | 26 | 1 | 2 | 21.77 |
| 6 | 42 | 1 | 1 | 28.67 | 18 | 31 | 1 | 2 | 23.98 |
| 7 | 27 | 2 | 1 | 29.62 | 19 | 40 | 1 | 2 | 23.41 |
| 8 | 35 | 2 | 1 | 26.53 | 20 | 41 | 1 | 2 | 22.97 |
| 9 | 55 | 1 | 1 | 30.56 | 21 | 31 | 2 | 2 | 20.18 |
| 10 | 53 | 1 | 1 | 29.52 | 22 | 31 | 1 | 2 | 20.62 |
| 11 | 54 | 1 | 1 | 24.80 | 23 | 32 | 2 | 2 | 21.87 |
| 12 | 53 | 2 | 1 | 28.29 | 24 | 43 | 2 | 2 | 23.23 |

It is desired to verify whether there is a significant effect of gender and intervention on the BMI. Use suitable ANOVA and draw conclusions.

**3.3** For the data given in Exercise 3.2, carry out two-factor ANOVA by taking age as a cofactor. Obtain the adjusted mean BMI for males and females separately.

**3.4** List various Post Hoc tests available, as a follow-up to one-way ANOVA in SPSS.

**3.5** The shoot length (SL) of a plant (in centimeters) measured at 25 days after sowing (DAS) has been measured under different experimental conditions marked as treatments A, B, C and D (Control). The data obtained from this experiment with 2 replicates is shown below.

| Treatment | A | D | B | B | A | D | A | C | C | D | C | B |
|-----------|----|----|----|----|----|----|----|----|----|----|----|----|
| SL | 15 | 19 | 20 | 21 | 17 | 20 | 15 | 19 | 22 | 20 | 19 | 21 |

Carry out ANOVA and check which of the treatments differ significantly with the control with respect to average shoot length.

**3.6** Use Kruskhal Wallis method of ANOVA for the data given in Exercise 3.2.

---

# Suggested Reading

1. Johnson, R.A., & Wichern, D.W. 2014. *Applied multivariate statistical analysis*, 6th ed. Pearson New International Edition.

2. Montgomery. D.C. 1997. *Design and Analysis of Experiments*. 4th ed. New York: John Wiley & Sons.

3. Daniel, W.W. 2009. *Biostatistics: Basic Concepts and Methodology for the Health Sciences*. 9th ed. John Wiley & Sons.

4. Zar, J.H. 2014. *Biostatistical Analysis*. 5th ed. Pearson New International Edition.

5. Sarma K.V.S. 2010. *Statistics Made Simple-Do it yourself on PC*. 2nd edition. Prentice Hall India.

# Chapter 4

# Multivariate Analysis of Variance (MANOVA)

| | | |
|---|---|---|
| 4.1 | Simultaneous ANOVA of Several Outcome Variables .......... | 73 |
| 4.2 | Test Procedure for MANOVA ................................. | 74 |
| 4.3 | Interpreting the Output ........................................ | 78 |
| 4.4 | MANOVA with Age as a Covariate ............................ | 81 |
| 4.5 | Using Age and BMI as Covariates ............................ | 84 |
| 4.6 | Theoretical Model for Prediction ............................. | 86 |
| | Summary ...................................................... | 88 |
| | Do it yourself (Exercises) .................................... | 88 |
| | Suggested Reading ............................................ | 90 |

Statistics is the grammar of science

Karl Pearson (1857 – 1936)

## 4.1 Simultaneous ANOVA of Several Outcome Variables

In the univariate ANOVA, there is a single response/outcome variable (Y), which is assumed to follow normal distribution with mean $\mu$ and a common variance $\sigma^2$ in all the groups (populations). It is also observed in Chapter-3 that normality of Y and homogeneity of variances are two fundamental assumptions, to be verified before applying ANOVA.

The Multivariate ANOVA (MANOVA) is an extension of univariate ANOVA for comparing the mean vectors of several independent groups, each vector having k ($\geqslant 2$) response variables. We may recall that in the case of only two groups, we use Hotelling's $T^2$ test for comparison. MANOVA can be applied in the following situations.

- The sample data contains n-subjects, and p-characteristics are measured on each subject.

- The sample subjects are segregated into g-independent groups (like stage of cancer or type of stimulant used) where $g \geqslant 3$.

- The mean vector on the p-characteristics from the $l^{th}$ group is $\mu_l$ which is a (1 x k) vector.

- The covariance matrix for the $l^{th}$ group is a (k x k) matrix $\Sigma_l$.

- It is assumed that $\Sigma_1 = \Sigma_2 = \ldots = \Sigma_g = \Sigma$ which means that the covariance matrices in all the groups are equal.

- The model for one-way MANOVA with g-groups and n-observations in each group is given by $X_{lj} = \mu + t_l + e_l$ for $l = 1, 2, \ldots, g$ and $j = 1, 2, \ldots, n_l$ where $n_l$ is the number of observations in the $l^{th}$ group such that $\sum_{l=1}^{g} n_l = n$ is the total sample size. $X_{lj}$ is called the *data matrix* for the $l^{th}$ group which is a matrix of size ($n_l$ x k). It is assumed that terms $e_{lj}$, called *error terms* are independent multivariate normally distributed with mean vector $\mathbf{0}$ and variance-covariance matrix $\Sigma$, denoted by $N_k(\mathbf{0}, \Sigma)$.

- $\mu$ is the overall mean from the entire data irrespective of the groups (classification) and $t_l$ is the effect of the $l^{th}$ group. In designed experiments, these groups are identified as treatments.

---

## 4.2   Test Procedure for MANOVA

The test procedure for one-way MANOVA is analogous to the univariate one-way ANOVA except that in each group we deal with *data matrices* instead of individual values. Each observed value $x_{lj}$ can be decomposed into three components viz., a) overall estimated mean vector ($\bar{x}$), b) estimated effect of the $l^{th}$ group and c) residual or error component.

This is expressed as

$$x_{lj} = \bar{x} + (\bar{x}_l - \bar{x}) + (x_{lj} - \bar{x}_l)$$

The total variation in all the data, expressed in terms of the Sum of Squares and Cross Product (SSCP) matrices can also be split into two components a) variation between groups and b) variation within groups given as follows:

$$\sum_{j=1}^{n_l}\sum_{l=1}^{g}(x_{lj}-\bar{x})(x_{lj}-\bar{x})' = \sum_{l=1}^{g}(\bar{x}_k-\bar{x})(\bar{x}_l-\bar{x})' + \sum_{j=1}^{n_l}\sum_{l=1}^{g}(x_{lj}-\bar{x}_l)(x_{lj}-\bar{x}_l)'$$

Define T = SSCP due to 'total' variation, B = SSCP due to 'between groups' variation and W = SSCP due to 'within groups' variation. Then the partitioning of the total sum of squares can be expressed as T = B + W.

The matrix B can also be expressed as $(n_1\text{-}1)\mathbf{S_1} + (n_2\text{-}1)\mathbf{S_2} + \ldots + (n_g\text{-}1)\mathbf{S_g}$ where $\mathbf{S_l}$ is the sample covariance matrix of data from the $l^{\text{th}}$ group.

In general if $\mathbf{S_{p\times p}}$ denotes the sample covariance matrix then the determinant of S is the generalized sample variance (discussed in Chapter 1) and produces a numerical value to measure the variation expressed by the matrix S. When p = 1 the quantity | S | (determinant of S) becomes the variance of a single variable. We use this concept and develop a test statistic similar to that of F-test used in one-way ANOVA. The MANOVA table appears as shown in Table 4.1

TABLE 4.1: General format for MANOVA

| Source of variation | df | Matrix of SSCP |
|---|---|---|
| Between groups | g-1 | $\mathbf{B} = \sum_{l=1}^{g}(\bar{x}_l-\bar{x})(\bar{x}_l-\bar{x})'$ |
| Within groups | $\sum_{l=1}^{g} n_l - g$ | $\mathbf{W} = \sum_{j=1}^{n_l}\sum_{l=1}^{g}(x_{lj}-\bar{x}_l)(x_{lj}-\bar{x}_l)'$ |
| **Total** | $\sum_{k=1}^{g} n_k - 1$ | $\mathbf{T} = \sum_{j=1}^{n_k}\sum_{k=1}^{g}(x_{kj}-\bar{x})(x_{kj}-\bar{x})'$ |

The test procedure is based on the *variance ratio* similar to the univariate ANOVA but the computations are based on matrices instead of scalar values. The commonly used method is the Wilk's lambda criterion discussed below.

**Wilk's Lambda criterion:**

This is based on the ratio of determinants of the generalized sample variance-covariance matrices **B** and **W** given by $\boldsymbol{\Lambda}^* = \dfrac{|\mathbf{W}|}{|\mathbf{B+W}|}$ which is re-

lated to Hotelling's $T^2$ statistic. Its value is measured by $\boldsymbol{\Lambda} = \sum_{m=1}^{r}(1+\lambda_m)^{-1}$

where $\lambda_1, \lambda_2, \ldots, \lambda_r$ are the 'm' *latent roots* or *eigen values* of the matrix $\mathbf{W^{-1}B}$.

We can find the eigen values and eigen vectors of any square matrix by using the *Real Statistics* MS-Excel Add-ins.

**Remark-1:**

An eigen value is a measure that counts the number of study variables which jointly explain a hidden or latent characteristic (usually unobservable) of data. If *eigen value* is 3, it means there are three variables (inter-correlated) in the data which together represent a characteristic like satisfaction or something such as 'being good'. These eigen values play a key role in problems of 'dimension reduction' in data.

It can be shown that for $g \geqslant 2$ and $p \geqslant 2$, the statistic

$$F = \left[ \frac{(n - g - k + 1)}{k} \right] \left[ \frac{(1 - \sqrt{\Lambda^*})}{\Lambda^*} \right]$$

follows Snedecor's F-distribution with $[k(g-1), (g-1)(n-g-k+1)]$ degrees of freedom.

The null hypothesis is rejected if the F-value is larger than the critical value at the chosen level of significance or when the p-value of the F-test is smaller than the error rate $\alpha$.

As alternatives to Wilk's lambda, there are other measures reported by statistical packages like SPSS and commonly adopted for testing the null hypothesis. These are based on the eigenvalues of the $\mathbf{W^{-1}B}$ matrix as given below.

- Pillai's trace $= \sum_{j=1}^{r} \frac{\lambda_j}{1 + \lambda_j}$ where $r = \min (p, g-1)$

- Hotelling's trace $= \sum_{j=1}^{r} \lambda_j$

- Roy's largest root $= \frac{\lambda_m}{1 + \lambda_m}$ where $\lambda_m$ is the largest eigenvalue of $\mathbf{W^{-1}B}$

It is established that in terms of the power of the test, Pillai's trace has the highest power followed by Wilk's lambda, Hotelling's trace and Roy's criterion in the order. The significance of the difference of mean vectors across the groups is reported by the p-value of each of these criteria.

**Box's M-test:**

Homogeneity of the covariance matrices across the groups can be tested with the help of Box's M-test. Rencher and Christensen (2012) made the following observations.

1. M-test is available in many software packages including SPSS.

2. Rejection of null hypothesis of 'equal covariance matrices' is not a serious problem when the number of observations is the same in each group.

It is also established that if the p-value of the M-test shows significance, then Pillai's trace will be a robust test which can be used in the place of Hotelling's trace or Wilk's lambda.

Consider the following illustration.

**Illustration 4.1** Reconsider the BMD data discussed in Illustration 3.1. The BMD profile is described by three variables *bmd_s, bmd_nf, bmd_gtf*. We now test the effect of i) *age group* and ii) *gender*, simultaneously on the BMD profile using MANOVA.

The following steps are required in SPSS.

1. Open the data file *bmddata.sav*

2. Click the options Analyse → General Linear Model → Multivariate

3. A dialogue box appears (see Figure 4.1) for selection of dependent variables and fixed factors. Let us take *bmd_s, bmd_nf, bmd_gtf* as dependent variables.

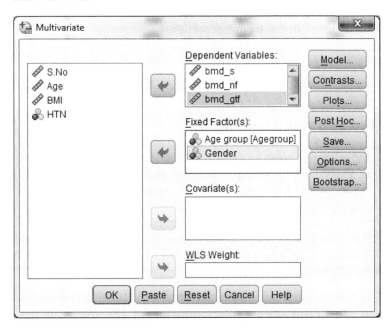

FIGURE 4.1: MANOVA-selection of variables.

4. The fixed factors shall be the *age group* (at 3 levels) and *gender* (two levels). By default, the *model* automatically assumes *full* factorial and hence the interaction term is included in the analysis.

5. Click on 'options' tab and select *Display Means*. The estimated marginal means will be produced by the model for *age group*, *gender* and their interaction. We will also get the estimated *overall* means for the profile variables.

6. Select *Homogeneity Tests* to verify the equality of covariance matrices among the groups. SPSS will report this in terms of Box's M test. If the p-value of the test is smaller than $\alpha$ (0.025 here) it means the assumption of homogeneity is *violated*.

7. Press the 'continue' button and as an option choose the *plots* tab. It is useful to examine the plot of mean vectors (profile plot) against *age group*, *gender* as well as their interaction, as additional information.

8. Press 'continue' and then press the OK button.

Several other options like *Save* and *Post Hoc* are beyond the present scope of discussion.

The entire sequence of steps can be saved as a syntax file for further use by clicking the *Paste* option. This completes the MANOVA procedure.

The output of MANOVA contains several tables and model indicates. These are discussed in the following section.

---

## 4.3   Interpreting the Output

The SPSS output contains several tables each depicting the results for all the profile variables, but we suppress a few and the results will be presented as text. Some tables of the output are rearranged for better understanding.

a) The *Box's M-test* for equality of covariance matrices shows M = 42.680 and p = 0.017 and hence the hypothesis of equal covariance matrices is not accepted. Still we proceed with the MANOVA test and use Pillai's trace for evaluation.

b) The multivariate test results are shown in Table 4.2. It can be seen that the Pillai's trace is not significant for *age group* but significant for *gender* (Trace = 0.234, F = 3.256 and p = 0.034). The interaction between *age group* and *gender* is also not significant (though Roy's test shows significance)

TABLE 4.2: Multivariate Tests of BMD profile

| Effect | Test | Value | F | Hypothesis df | Error df | Sig. |
|--------|------|-------|---|---------------|----------|------|
| Intercept | Pillai's Trace | 0.981 | 539.513[b] | 3 | 32 | 0.000 |
| | Wilks' Lambda | 0.019 | 539.513[b] | 3 | 32 | 0.000 |
| | Hotelling's Trace | 50.579 | 539.513[b] | 3 | 32 | 0.000 |
| | Roy's Largest Root | 50.579 | 539.513[b] | 3 | 32 | 0.000 |
| Age group | Pillai's Trace | 0.269 | 1.711 | 6 | 66 | 0.132 |
| | Wilks' Lambda | 0.747 | 1.672[b] | 6 | 64 | 0.142 |
| | Hotelling's Trace | 0.316 | 1.633 | 6 | 62 | 0.153 |
| | Roy's Largest Root | 0.211 | 2.323[c] | 3 | 33 | 0.093 |
| Gender | Pillai's Trace | 0.234 | 3.256[b] | 3 | 32 | 0.034 |
| | Wilks' Lambda | 0.766 | 3.256[b] | 3 | 32 | 0.034 |
| | Hotelling's Trace | 0.305 | 3.256[b] | 3 | 32 | 0.034 |
| | Roy's Largest Root | 0.305 | 3.256[b] | 3 | 32 | 0.034 |
| Age group * Gender | Pillai's Trace | 0.280 | 1.788 | 6 | 66 | 0.115 |
| | Wilks' Lambda | 0.735 | 1.776[b] | 6 | 64 | 0.118 |
| | Hotelling's Trace | 0.341 | 1.763 | 6 | 62 | 0.122 |
| | Roy's Largest Root | 0.268 | 2.945[c] | 3 | 33 | 0.047 |

a. Design: Intercept + Age group + Gender + Age group * Gender.
b. Exact statistic.
c. The statistic is an upper bound on F that yields a lower bound on the significance level.

c) Since the multivariate test is significant with at least one factor (*gender*), we proceed to perform univariate tests with respect to *age group* and *gender* on *bmd_s, bmd_nf, bmd_gtf* separately. These results appear in the table titled *Tests of Between-Subjects Effects* in the SPSS output and are shown in Table 4.3.

d) It follows that *bmd_s* and *bmd_gtf* differ significantly with *gender* (p = 0.018 and p = 0.009 respectively).

TABLE 4.3: Tests of between-subjects effects (univariate ANOVA)

| Source | Dependent Variable | Type III Sum of Squares | df | Mean Square | F | p-value |
|---|---|---|---|---|---|---|
| Corrected Model | bmd_s | 0.163[a] | 5 | 0.033 | 2.341 | 0.063 |
| | bmd_nf | 0.087[b] | 5 | 0.017 | 1.655 | 0.172 |
| | bmd_gtf | 0.114[c] | 5 | 0.023 | 2.492 | 0.050 |
| Intercept | bmd_s | 22.741 | 1 | 22.741 | 1634.295 | 0.000 |
| | bmd_nf | 15.634 | 1 | 15.634 | 1486.993 | 0.000 |
| | bmd_gtf | 11.469 | 1 | 11.469 | 1253.283 | 0.000 |
| Age group | bmd_s | 0.013 | 2 | 0.006 | 0.462 | 0.634 |
| | bmd_nf | 0.038 | 2 | 0.019 | 1.813 | 0.179 |
| | bmd_gtf | 0.046 | 2 | 0.023 | 2.514 | 0.096 |
| Gender | bmd_s | 0.086 | 1 | 0.086 | 6.208 | 0.018 |
| | bmd_nf | 0.037 | 1 | 0.037 | 3.517 | 0.069 |
| | bmd_gtf | 0.070 | 1 | 0.070 | 7.674 | 0.009 |
| Age group * Gender | bmd_s | 0.041 | 2 | 0.021 | 1.474 | 0.243 |
| | bmd_nf | 0.004 | 2 | 0.002 | 0.182 | 0.835 |
| | bmd_gtf | 0.011 | 2 | 0.005 | 0.588 | 0.561 |
| Error | bmd_s | 0.473 | 34 | 0.014 | | |
| | bmd_nf | 0.357 | 34 | 0.011 | | |
| | bmd_gtf | 0.311 | 34 | 0.009 | | |
| Total | bmd_s | 29.669 | 40 | | | |
| | bmd_nf | 20.482 | 40 | | | |
| | bmd_gtf | 14.653 | 40 | | | |
| Corrected Total | bmd_s | 0.636 | 39 | | | |
| | bmd_nf | 0.444 | 39 | | | |
| | bmd_gtf | 0.425 | 39 | | | |

a. R Squared = 0.256 (Adjusted R Squared = 0.147).
b. R Squared = 0.196 (Adjusted R Squared = 0.077).
c. R Squared = 0.268 (Adjusted R Squared = 0.161).

e) Since *gender* has shown a significant effect on the BMD profile, it is worth reporting the estimated marginal mean values under the influence of the MANOVA model as shown in Table 4.4.

In all the cases, male patients show a higher BMD than females but *bmd_s* and *bmd_gtf* have a significant difference (from Table 4.3) due to gender.

TABLE 4.4: Estimated marginal means due to gender

| Dependent Variable | Gender | Mean | Std. Error | 95% Confidence Interval | |
|---|---|---|---|---|---|
| | | | | Lower Bound | Upper Bound |
| bmd_s | Male | 0.921 | 0.037 | 0.847 | 0.995 |
| | Female | 0.814 | 0.022 | 0.769 | 0.860 |
| bmd_nf | Male | 0.754 | 0.032 | 0.690 | 0.819 |
| | Female | 0.684 | 0.019 | 0.645 | 0.724 |
| bmd_gtf | Male | 0.664 | 0.030 | 0.604 | 0.725 |
| | Female | 0.568 | 0.018 | 0.531 | 0.605 |

In the following section we see the advantage of having *actual age* as a covariate instead of using *age group* as a factor.

## 4.4 MANOVA with Age as a Covariate

This can be done by removing *age group* from the factor list and introducing the variable *age* as a covariate and repeating the entire exercise again. We get the results after choosing the options as shown in Figure 4.2.

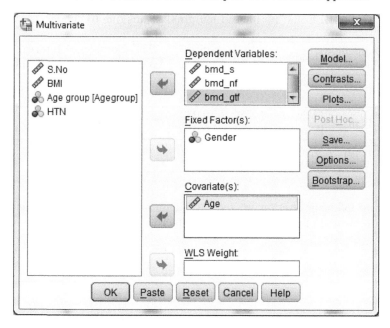

FIGURE 4.2: MANOVA-selection of covariate.

Even though, *gender* leads to only two groups we use MANOVA since *age* is a covariate that fits into a General Linear Model. Without this covariate, a Hotelling's $T^2$ would be the appropriate test.

We make the following observations:

a) The multivariate test with respect to *age* has Pillai's trace $= 0.205$, F $= 3.007$ and p $= 0.043$ and hence *age* has a significant effect on the BMD profile.

b) With respect to *gender* we get Pillai's trace $= 0.248$, F $= 3.848$ and p $= 0.018$ showing that *gender* is a significant factor after adjusting for *age*.

c) From the table 'Tests Between subjects' of SPSS output we notice that both *age* and *gender* show a significant effect on the three variables (p $< 0.05$) as displayed in Table 4.5. We may recall that when *age* was not a covariate, *bmd_nf* did not show any influence of *gender*!

TABLE 4.5: Univariate ANOVA with age as covariate

| Source | Dependent Variable | F | p-value |
|--------|--------------------|----|---------|
| | bmd_s | 5.019 | 0.031 |
| Age | bmd_nf | 9.139 | 0.005 |
| | bmd_gtf | 5.022 | 0.031 |
| | bmd_s | 7.682 | 0.009 |
| Gender | bmd_nf | 4.361 | 0.044 |
| | bmd_gtf | 8.990 | 0.005 |

d) The estimated marginal mean of BMD after adjusting for *age* can be obtained by choosing the *Display Means* option and we get the mean values along with confidence intervals as shown in Table 4.6. The method works out the estimated Y values after fixing the covariate at the average of the factor values.

TABLE 4.6: Estimated marginal means due to gender adjusted for age

| Dependent Variable | Gender | Mean | Std. Error | 95% Confidence Interval | |
|--------------------|--------|------|------------|------------|------------|
| | | | | Lower Bound | Upper Bound |
| bmd_s | Male | $0.934^a$ | 0.035 | 0.864 | 1.005 |
| | Female | $0.821^a$ | 0.021 | 0.778 | 0.864 |
| bmd_nf | Male | $0.759^a$ | 0.029 | 0.701 | 0.818 |
| | Female | $0.688^a$ | 0.018 | 0.652 | 0.724 |
| bmd_gtf | Male | $0.668^a$ | 0.028 | 0.611 | 0.725 |
| | Female | $0.569^a$ | 0.017 | 0.534 | 0.604 |

a. Covariates appearing in the model are evaluated at the following values: *age* = 35.50.

It is easy to see that these means are different from those presented in Table 4.4 where *age* was not a covariate.

**Remark-2:**

Since *age* is a continuous variable it is not possible to define an interaction term. That is the reason we did not get a two-way table of means as we got for the *age group*. Only categorical factors and their combinations appear in cross tabulated means.

Suppose we have more than one continuous covariate that can influence the outcome variable. We can accommodate all such covariates into the model and improve the test procedure because the outcome will be adjusted for these covariates.

**Remark-3:**

Before introducing a variable as covariate into the model, it is important to verify that it has a *non-zero correlation* with the outcome variable. An uncorrelated covariate conveys nothing. For instance in the case of *age*, the three outcome variables had correlation coefficients -0.275, -0.398 and -0.268 respectively.

In the following section we study the effect of including *age* and BMI as two covariates.

---

## 4.5    Using Age and BMI as Covariates

We can include more than one covariate also into the model and check their effect on the BMD profile. The variable BMI can also be used as a covariate for which the correlation coefficients with the profile variables are 0.349, 0.344 and 0.384 respectively.

We only need to send both *age* and BMI into the covariate box and run with the remaining options unchanged.

The following results are obtained.

a) The model has $R^2$= 0.351, 0.374 and 0.394 for *bmd_s*, *bmd_nf* and *bmd_gtf* respectively.

b) The multivariate test with respect to *age* has Pillai's trace = 0.261, F = 4.000 and p = 0.015 and hence *age* has a significant effect on the BMD profile.

c) With respect to BMI we get Pillai's trace = 0.202, F = 2.868 and p = 0.051 showing that BMI has a significant effect after adjusting for *age*.

d) With respect to *gender* we get Pillai's trace = 0.272, F = 4.231 and p = 0.012 showing that *gender* is a significant factor after adjusting for *age* and BMI

e) The univariate ANOVA for *age*, BMI and *gender* shows a significant effect on the individual variables ($p < 0.05$) as displayed in Table 4.7.

TABLE 4.7: Univariate ANOVA with covariates age and BMI

| Source | Dependent Variable | F | p-value |
|--------|-------------------|-----|---------|
|        | bmd_s  | 7.417  | 0.011 |
| Age    | bmd_nf | 12.663 | 0.001 |
|        | bmd_gtf | 7.662 | 0.009 |
|        | bmd_s  | 6.428  | 0.016 |
| BMI    | bmd_nf | 7.307  | 0.010 |
|        | bmd_gtf | 7.662 | 0.006 |
|        | bmd_s  | 8.290  | 0.007 |
| Gender | bmd_nf | 4.688  | 0.037 |
|        | bmd_gtf | 10.213 | 0.003 |

f) It may be observed from the complete output that the $R^2$ has been increasing as we introduce more factors or covariates, indicating that the model is adequate.

g) The estimated marginal mean of the BMD profile after adjusting for *age* and BMI can be obtained by choosing the *Display Means* option. The mean values along with confidence intervals are shown in Table 4.8. The method of adjustment to covariates works out the estimated Y values after fixing *age* at mean = 35.50 years and average BMI at 20.55.

TABLE 4.8: Estimated marginal means due to gender adjusted for age and BMI

| Dependent Variable | Gender | Mean | Std. Error | 95% Confidence Interval | |
|--------------------|--------|------|------------|-------------|-------------|
|                    |        |      |            | Lower Bound | Upper Bound |
| bmd_s   | Male   | 0.932[a] | 0.032 | 0.866 | 0.998 |
|         | Female | 0.822[a] | 0.020 | 0.781 | 0.862 |
| bmd_nf  | Male   | 0.757[a] | 0.027 | 0.703 | 0.811 |
|         | Female | 0.689[a] | 0.016 | 0.656 | 0.722 |
| bmd_gtf | Male   | 0.666[a] | 0.026 | 0.614 | 0.718 |
|         | Female | 0.570[a] | 0.016 | 0.538 | 0.602 |

a. Covariates appearing in the model are evaluated at the following values: *age* = 35.50, BMI = 20.55.

**Remark-4:**

We may use the analytical results of MANOVA to predict the BMD profile of a new individual, by using information on the predictive factors used in the model. The accuracy of prediction will be more when the MANOVA model is a good fit and the corresponding $R^2$ values of the univariate ANOVA are high. One should note that predicting the BMD value is only in terms of the average value at a given *age*, BMI and *gender* and does not mean the exact value for an individual, but it is an average!

In the following section we discuss methods for decision making for the prediction of BMD of a new individual.

---

## 4.6    Theoretical Model for Prediction

At the end of the study we arrive at a model (formula) to predict the BMD profile of an individual, in terms of *age*, BMI and *gender*. This is obtained by choosing the SPSS option *parameter estimates*. The output appears as shown in Table 4.9.

TABLE 4.9: Parameter estimates of BMD using covariates

| Dependent Variable | Parameter | B | Std. Error | t | Sig. | 95% CI |
|---|---|---|---|---|---|---|
| | Intercept | 0.727 | 0.115 | 6.332 | 0.000 | (0.494, 0.960) |
| | Age | -0.005 | 0.002 | -2.673 | 0.011 | (-0.008, -0.001) |
| bmd_s | BMI | 0.013 | 0.005 | 2.535 | 0.016 | (0.003, 0.023) |
| | [Gender=1] | 0.110 | 0.038 | 2.879 | 0.007 | (0.033, 0.188) |
| | [Gender=2] | $0^a$ | - | - | - | - |
| | Intercept | 0.643 | 0.094 | 6.817 | 0.000 | (0.452, 0.834) |
| | Age | -0.005 | 0.001 | -3.559 | 0.001 | (-0.008, -0.002) |
| bmd_nf | BMI | 0.011 | 0.004 | 2.703 | 0.010 | (0.003, 0.019) |
| | [Gender=1] | 0.068 | 0.031 | 2.165 | 0.037 | (0.004, 0.132) |
| | [Gender=2] | $0^a$ | - | - | - | - |
| | Intercept | 0.471 | 0.091 | 5.194 | 0.000 | (0.287, 0.655) |
| | Age | -0.004 | 0.001 | -2.768 | 0.009 | (-0.007, -0.001) |
| bmd_gtf | BMI | 0.011 | 0.004 | 2.890 | 0.006 | (0.003, 0.019) |
| | [Gender=1] | 0.096 | 0.030 | 3.182 | 0.003 | (0.035, 0.157) |
| | [Gender=2] | $0^a$ | - | - | - | - |

a. This parameter is set to zero because it is redundant.

The output gives the estimated marginal contribution of the factors and covariates indicated in column B. It is called the *regression coefficient* of the factor. It represents the marginal increase (or decrease depending on the sign) in the outcome due to a unit increase in the factor.

The output also provides the standard error of this estimate. A t-test is performed to test whether the regression coefficient is significantly different from zero (hypothetical value). We find that all the regression coefficients are significant ($p < 0.05$).

In this case the general form of the model is

$$bmd = \text{intercept} + b1 * age + b2 * BMI + b3 * gender$$

The following prediction formulas can be arrived at from Table 4.9.

1. $bmd\_s = 0.727 - 0.005 * age + 0.013 * BMI + 0.110 * gender$

2. $bmd\_nf = 0.643 - 0.005 * age + 0.011 * BMI + 0.068 * gender$

3. $bmd\_gtf = 0.471 - 0.004 * age + 0.011 * BMI + 0.096 * gender$

Suppose a female patient is presenting with $age = 35$ years and BMI $= 22.0$. Then in the equation for $bmd\_s$, substituting $gender = 0$, the predicted $bmd\_s$ will be

$$bmd\_s = 0.727 - 0.005 * 35 + 0.013 * 22 + 0.110 * 0$$

and this gives 0.825.

(Even though we have used the code '2' for female, the model converts '2' to '0' by creating a *dummy variable*. So we have to use '0'.)

Figure 4.3 shows a simple MS-Excel template to perform the above calculations. The inputs can be changed and the corresponding predicted values can be automatically obtained.

| F4 | | | $\times$ $\checkmark$ | $f_x$ | {=SUMPRODUCT($B$3:$E$3,B4:E4)} | |
|---|---|---|---|---|---|---|
| | A | B | C | D | E | F |
| 1 | | Intercept | Age | BMI | Gender | Predicited |
| 2 | | | | | ( Male=1, Female=0) | BMD |
| 3 | Inputs → | 1 | 35 | 22 | 0 | |
| 4 | bmd_s | 0.727 | -0.005 | 0.013 | 0.110 | 0.842 |
| 5 | bmd_nf | 0.643 | -0.005 | 0.011 | 0.068 | 0.708 |
| 6 | bmd_gtf | 0.471 | -0.004 | 0.011 | 0.096 | 0.588 |
| 7 | | | | | | |
| 8 | **Note:** Inputs change from case to case. For intercept always '1' | | | | | |

FIGURE 4.3: MS-Excel template to predict BMD values.

Similarly for the same patient the predicted $bmd\_nf$ will be 0.720 and $bmd\_gtf$ will be 0.585 (from the other two equations).

With the same *age* and BMI if the patient happened to be male, we have to put '1' for input of *gender*. This gives predicted $bmd\_s = 0.948$, $bmd\_nf = 0.778$ and $bmd\_gtf = 0.669$.

Thus MANOVA helps in building predictive models for multivariate profiles using linear regression.

---

## Summary

MANOVA is a procedure to find whether the mean values of a panel of variables differ significantly among three or more groups of subjects. When the multivariate test indicated by Hotelling's trace or Pillai's trace is found to be significant, we may consider the difference among the groups as important and proceed to inspect each variable of the panel for significance. This is done by performing one-way ANOVA. SPSS has a module to include continuous cofactors (covariates) also into the model and derive the mean values of the outcome variables, after adjusting for the effect of covariates. The user of the procedure, however, has to ensure the validity of the assumptions such as normality of the data and homogeneity of covariance matrices.

---

## Do it yourself (Exercises)

4.1 The following data refers to selected parameters observed in an anthropometric study. For each subject (person) the Physical Activity (PA) Body Mass Index (BMI), Heart Rate (HR), Total Count (TC) of WBC per cubic mm of blood and High Density Lipoproteins (HDL) md/dl have been measured. The data also contains information on *age* and *gender* of each person. A sample of 22 records is given below.

   a) Find the covariance matrix of the profile BMI, HR, TC, HDC.

   b) Perform MANOVA and examine the effect of physical activity on the health profile in terms of the 4 variables BMI, HR, TC, HDC.

   c) How would the results change if *age* is included as a covariate and only *gender* as a factor?

| S.No | Age | Gender | PA | BMI | HR | TC | HDL |
|------|-----|--------|----|----|----|----|----|
| 1 | 28 | Male | Mild | 20.48 | 82 | 95.67 | 46.00 |
| 2 | 30 | Male | Moderate | 23.07 | 75 | 95.00 | 48.48 |
| 3 | 50 | Male | Moderate | 24.22 | 72 | 123.34 | 47.78 |
| 4 | 28 | Female | Mild | 20.20 | 77 | 120.00 | 50.55 |
| 5 | 33 | Male | Moderate | 21.19 | 75 | 134.21 | 49.57 |
| 6 | 26 | Female | Mild | 21.10 | 77 | 96.67 | 47.01 |
| 7 | 44 | Female | Mild | 23.74 | 68 | 130.23 | 45.00 |
| 8 | 40 | Female | Moderate | 21.34 | 65 | 121.22 | 54.32 |
| 9 | 38 | Female | Moderate | 22.96 | 70 | 120.09 | 46.77 |
| 10 | 45 | Female | Mild | 20.13 | 75 | 100.00 | 47.21 |
| 11 | 28 | Female | Moderate | 20.96 | 72 | 88.89 | 53.23 |
| 12 | 27 | Female | Mild | 21.19 | 72 | 95.79 | 48.15 |
| 13 | 26 | Male | Moderate | 21.88 | 76 | 100.70 | 47.00 |
| 14 | 25 | Female | Moderate | 22.94 | 72 | 90.33 | 49.57 |
| 15 | 31 | Male | Moderate | 24.51 | 74 | 122.23 | 47.32 |
| 16 | 28 | Female | Mild | 23.92 | 69 | 111.23 | 49.50 |
| 17 | 54 | Male | Mild | 21.80 | 77 | 121.23 | 43.34 |
| 18 | 55 | Female | Mild | 23.04 | 75 | 152.00 | 46.12 |
| 19 | 55 | Female | Mild | 22.89 | 74 | 137.78 | 48.98 |
| 20 | 29 | Male | Moderate | 23.19 | 75 | 86.66 | 47.00 |
| 21 | 40 | Male | Moderate | 23.88 | 70 | 117.77 | 49.21 |
| 22 | 34 | Female | Mild | 23.68 | 74 | 124.00 | 48.00 |

(Data Courtesy: Dr. Kanala Kodanda Reddy, Department of Anthropology, Sri Venkateswara University, Tirupati.)

**4.2** Haemogram and *lipid profile* are two examples where several related parameters are simultaneously observed for the same patient. Collect a sample data on the lipid profile and study the following.

   a) Mean values and covariance matrix.
      (Hint: use MS-Excel Real Statistics)

   b) Plot the mean values of profile variables.

   c) Correlation coefficients among lipid variables.

**4.3** What are the SPSS options available to create a custom design of the MANOVA model? By default it is taken as full-factorial for all factors and their interactions. How do you choose specific factors and a few interactions?

**4.4** The goodness of the MANOVA model can be understood in terms of $R^2$ value or adjusted $R^2$ value. When the $R^2$ is apparently low, we may try including other relevant factors or include covariates. Create your own case study and by trial and error, find out the factors and covariates that fit well to the MANOVA data.

**4.5** As a post hoc to MANOVA, obtain a plot of the estimated marginal means due to each factor. Use MS-Excel to create a plot of the BMD profile with respect to *age group* and *gender*.
(Hint: Use the table of estimated means and standard errors obtained from the SPSS option *age group * gender*.)

---

# Suggested Reading

1. Johnson, R. A., & Wichern, D. W. 2014. *Applied multivariate statistical analysis*, 6$^{th}$ ed. Pearson New International Edition.

2. Alvin C.Rencher, William F. Christensen 2012. *Methods of Multivariate Analysis*. 3$^{rd}$ ed. Brigham Young University: John Wiley & Sons.

3. Huberty, C.J. and Olejnik, S. 2006. *Applied MANOVA and Discriminant Analysis*. 2$^{nd}$ ed. John Wiley & Sons.

# Chapter 5

## Analysis of Repeated Measures Data

5.1    Experiments with Repeated Measures ..........................    91
5.2    RM ANOVA Using SPSS .......................................    95
5.3    RM ANOVA Using MedCalc ..................................    99
5.4    RM ANOVA with One Grouping Factor ......................   100
5.5    Profile Analysis ................................................   107
       Summary ......................................................   112
       Do it yourself (Exercises) ....................................   113
       Suggested Reading ............................................   114

Nature is written in mathematical language.

Galileo Galilei (1564 – 1642)

## 5.1  Experiments with Repeated Measures

Often we come across situations where the outcome of an experiment is measured on the same subject (like plant, animal or a human being) under different conditions or at different time points. The design of such an experiment is called a *repeated measures (RM) design*.

The univariate paired comparison test is one such example where the response to a treatment is measured before and after giving a treatment. Since the same measurement is made on the same individual under different conditions, the data values cannot be considered as independent.

Here are some instances where the study fits into an RM design.

a) The Quality of Life (QOL) after physiotherapy for post-surgical patients obtained three times viz., before treatment, at time of discharge and after a follow-up period.

b) Food is a major source to meet the nutritional needs of an individual. *Cassia tora* is one such nutrient which is under consumption among the Kurichiya tribe of Kerala, India (Himavanth Reddy Kambalachenu *et al.* (2018)). As part of the study the research team measured the High Density Lipoproteins (HDL) 6 times with a gap of 3 months using a crossover design. This experiment is a case of RM design.

c) The BMD observed at three locations viz., *bmd_s*, *bmd_nf* and *bmd_gtf* for each patient also forms a case of RM design. We can compare the changes due to location where the measurement was taken and also check the effect of a factor like gender or age on the BMD.

Let $X_1$, $X_2$ and $X_3$ denote three measurements taken on each individual at three instances. Define three new variables, $d_1 = (X_1 - X_2)$, $d_2 = (X_1 - X_3)$ and $d_3 = (X_2 - X_3)$ as the paired differences among the values. We may then hypothesize that $d_1 = d_2 = d_3 = 0$ or equivalently in terms of the mean values and we may take $H_0$: $\mu_1 = \mu_2 = \mu_3 = 0$ as the null hypothesis whose truth can be verified in the light of the sample data.

Since there are 3 conditions in which the means are being compared, we can use one-way ANOVA which ultimately performs the F-test. One of the assumptions of ANOVA is that $d_1$, $d_2$ and $d_3$ are independent, which is not true here. In fact they are correlated within each subject. Therefore we use a test known as *Repeated Measures (RM) ANOVA*. The method is based on a multivariate comparison of means using a test similar to Hotelling's trace for *within subjects* comparison.

Further, it is assumed that the data possesses a property called *sphericity* to mean that the variances of the pairwise differences, $d_1$, $d_2$ and $d_3$ are all equal by hypothesis viz., $V(d_1) = V(d_2) = V(d_3)$. The statistical test for the significance of sphericity is known as *Mauchly's W Test*. When the test result shows significance ($p < 0.05$), we conclude that the sphericity condition is violated. In such cases there will be a loss of power of the F-test in the RM ANOVA.

We may recall such a situation in the univariate two-sample Student's t-test, where the Levene's test is used to confirm equality of variances. If this test shows significance, it means that the variances are not equal and hence the degrees of freedom are adjusted to provide correct p-value for the test.

When the Mauchly's test of sphericity shows significance, it means the sphericity assumption is violated. In that case, the degrees of freedom are adjusted before evaluating the p-value. SPSS reports a statistic called $\varepsilon$ (epsilon) which is a measure of extent of departure from sphericity.

When $\varepsilon = 1$ the sphericity condition holds well and all the F-tests regarding the components of the RM factor hold well. RM factor is a categorical variable indicating the different contexts where measurements are observed repeatedly.

When $\varepsilon < 1$, we need to choose a test procedure that takes into account the adjusted degrees of freedom. The value of $\varepsilon$ is evaluated by three methods viz., a) Greenhouse-Geisser method, b) Huynh-Feldt method and c) method of lower bound.

Comparison of the merits of each method are beyond the scope of the present discussion. However a guiding rule suggested by Rencher and Christensen (2012) is as follows.

a) If $\varepsilon > 1$, no correction required.

b) If $\varepsilon > 0.75$, use the Hyunh-Feldt method.

c) If $\varepsilon \leqslant 0.75$, the Greenhouse-Geisser method is recommended.

The F-test due to the RM factor is then evaluated by these three methods and we have to select the one based on the $\varepsilon$ criterion.

Consider the following illustration.

**Illustration 5.1** The Quality of Life (QOL) scores (maximum = 100) of post-surgical patients undergoing physiotherapy are obtained by a researcher.

The patients are divided into a study group and control group and the QOL is measured at three points of time viz., before surgery (QOL_1), at the time of discharge (QOL_2) and after physiotherapy for 3 months (QOL_3).

The following codes are used:

Group: 1 = Graded Aerobic exercises for 3 months (Study Group), 2 = Non-graded Aerobic exercises for 3 months (Control Group). Gender: 1 = Male, 2 = Female

The data given in Table 5.1 shows the scores of 20 patients from the study. The analysis is however carried out on the complete data.

We wish to test whether there is any significant difference in the mean QOL during the study period.

TABLE 5.1: Sample value of QOL data and differences

| S.No | Group | Age | Gender | QOL_1 | QOL_2 | QOL_3 |
|------|-------|-----|--------|-------|-------|-------|
| 1    | 2     | 21  | 1      | 62    | 61    | 65    |
| 2    | 2     | 25  | 1      | 61    | 60    | 63    |
| 3    | 2     | 26  | 2      | 60    | 59    | 63    |
| 4    | 2     | 26  | 2      | 62    | 60    | 63    |
| 5    | 2     | 26  | 1      | 62    | 61    | 64    |
| 6    | 2     | 27  | 1      | 61    | 60    | 64    |
| 7    | 2     | 28  | 2      | 62    | 60    | 65    |
| 8    | 2     | 28  | 1      | 63    | 59    | 67    |
| 9    | 2     | 29  | 1      | 65    | 62    | 68    |
| 10   | 2     | 30  | 1      | 58    | 55    | 62    |
| 11   | 1     | 30  | 1      | 55    | 54    | 56    |
| 12   | 1     | 30  | 1      | 57    | 54    | 58    |
| 13   | 1     | 30  | 1      | 58    | 56    | 60    |
| 14   | 1     | 30  | 2      | 60    | 58    | 59    |
| 15   | 1     | 29  | 2      | 58    | 56    | 60    |
| 16   | 1     | 25  | 2      | 60    | 58    | 59    |
| 17   | 1     | 26  | 1      | 58    | 56    | 55    |
| 18   | 1     | 26  | 2      | 61    | 60    | 59    |
| 19   | 1     | 32  | 2      | 59    | 58    | 57    |
| 20   | 1     | 34  | 2      | 60    | 59    | 58    |

*(Data courtesy: Dr Abha Chandra, Department of CT Surgery,*
*Sri Venkateswara Institute of Medical Sciences (SVIMS), Tirupati.)*

**Analysis:**

Let us compute the three possible differences $d_1 = (QOL\_2 - QOL\_1)$, $d_2 = (QOL\_3 - QOL\_1)$ and $d_3 = (QOL\_3 - QOl\_2)$. The mean and S.D of $d_1$, $d_2$ and $d_3$ convey the pattern of change over the three time periods. These calculations are also shown below for the data given in Table 5.1.

| S.No | 1  | 2  | 3  | 4  | 5  | 6  | 7  | 8  | 9  | 10 |
|------|----|----|----|----|----|----|----|----|----|----|
| d1   | -1 | -1 | -1 | -2 | -1 | -1 | -2 | -4 | -3 | -3 |
| d2   | 3  | 2  | 3  | 1  | 2  | 3  | 3  | 4  | 3  | 4  |
| d3   | 4  | 3  | 4  | 3  | 3  | 4  | 5  | 8  | 6  | 7  |

| S.No | 11 | 12 | 13 | 14 | 15 | 16 | 17 | 18 | 19 | 20 |
|------|----|----|----|----|----|----|----|----|----|----|
| d1   | -1 | -3 | -2 | -2 | -2 | -2 | -2 | -1 | -1 | -1 |
| d2   | 1  | 1  | 2  | -1 | 2  | -1 | -3 | -2 | -2 | -2 |
| d3   | 2  | 4  | 4  | 1  | 4  | 1  | -1 | -1 | -1 | -1 |

The mean $\pm$ S.D of $d_1$, $d_2$ and $d_3$ are -1.8 $\pm$ 0.894, 1.15 $\pm$ 2.207 and 2.95 $\pm$ 2.645 respectively.

We first find that the mean difference is not the same for the three instances with $d_3$ showing a higher difference than the other two.

We wish to test the hypothesis that the mean QOL remains the same at the three instances.

In the section below we discuss method of working with RMANOVA using SPSS.

## 5.2   RM ANOVA Using SPSS

We can perform RM ANOVA with SPSS following the menu sequence: Analyze $\rightarrow$ General Linear Model $\rightarrow$ Repeated Measures.

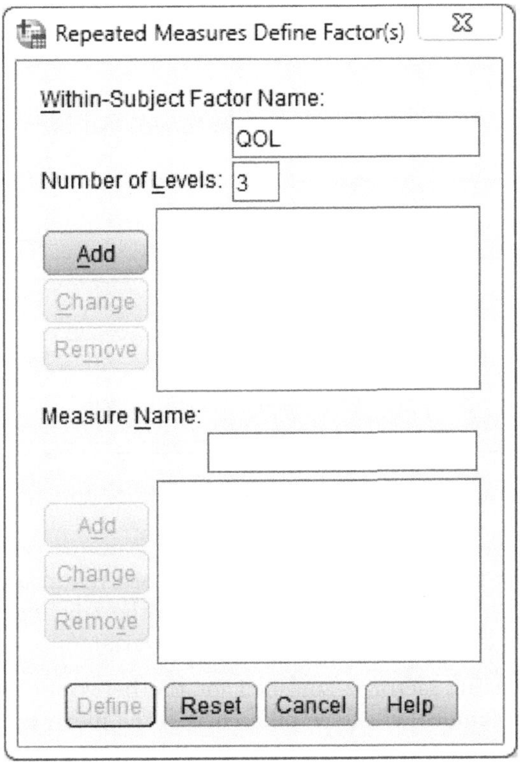

FIGURE 5.1: Defining the levels of the RM factor.

The procedure defines a new variable, called *within subjects factor*, displayed by default as *factor1* and seeks the number of levels at which the factor1 is observed. We can rename this new variable suitably for better understanding.

For instance in this context, we might rename factor1 as QOL and specify the number of levels as 3. Then click the *add* button to create this new factor. Then press the *define* button. This is shown in Figure 5.1.

This displays another window in which the variables are defined for the three levels, as shown in Figure 5.2.

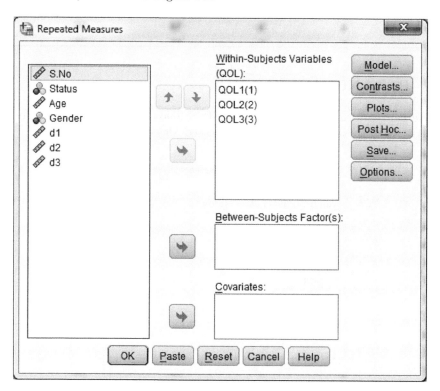

FIGURE 5.2: Naming the RM factor.

The input box labelled *between subjects factor(s)* is meant for specifying the grouping factor like *status, gender* etc., and the label covariates indicate provision to include continuous variable(s) as covariate(s).

When no grouping factor is specified for the between-subjects effect, we can still test the significance of the difference in the mean values of the three RM variables. It means, we test the hypothesis $H_0$: $\mu_1 = \mu_2 = \mu_3$ where $\mu_1 =$ Mean of $QOL\_1$, $\mu_2 =$ Mean of $QOL\_2$ and $\mu_3 =$ Mean of $QOL\_3$, without any grouping factor.

By pressing the OK button at this stage, we get the following output.

1. The factor QOL is a multivariate profile and we get Hotelling's trace = 3.140 with F = 167.079 and p < 0.001. Hence the difference among the means is significant, thereby rejecting the hypothesis of equality of means.

2. The Mauchly's W test shows significance with W = 0.278 and p < 0.001. Hence the sphericity condition is violated. It means the variances of the differences in the three variables are not the same. Hence the adjusted F-ratios are to be considered which are automatically given in Table 5.3.

TABLE 5.2: Mauchly's test of sphericity

| Within Subjects Effect | Mauchly's W | Approx. Chi-Square | df | Sig. | Epsilon | | |
| --- | --- | --- | --- | --- | --- | --- | --- |
| | | | | | Greenhouse-Geisser | Huynh-Feldt | Lower-bound |
| QOL | 0.278 | 125.413 | 2 | 0.000 | 0.581 | 0.583 | 0.500 |

3. The tests of within-subject effects for QOL are shown in Table 5.3 with F-ratio produced with the three types of corrections for degrees of freedom when sphericity is not assumed. Since W < 0.75 we can use the Greenhouse-Geisser method. This gives new degrees of freedom as 2*0.581 = 1.162 with F = 82.440 and p < 0.001. Hence we consider that the three repeated measures of QOL do differ significantly.

TABLE 5.3: Tests of within-subjects effects of QOL

| Source | Method | Type III Sum of Squares | df | Mean Square | F | Sig. |
| --- | --- | --- | --- | --- | --- | --- |
| QOL | Sphericity assumed | 302.607 | 2 | 151.303 | 82.44 | <0.001 |
| | Greenhouse-Geisser | 302.607 | 1.162 | 260.527 | 82.44 | <0.001 |
| | Huynh-Feldt | 302.607 | 1.167 | 259.362 | 82.44 | <0.001 |
| | Lower-bound | 302.607 | 1 | 302.607 | 82.44 | <0.001 |
| Error (QOL) | Sphericity assumed | 363.393 | 198 | 1.835 | | |
| | Greenhouse-Geisser | 363.393 | 114.99 | 3.160 | | |
| | Huynh-Feldt | 363.393 | 115.507 | 3.146 | | |
| | Lower-bound | 363.393 | 99 | 3.671 | | |

4. The next question is about the type of relationship among the mean of QOL at the three repeated measurements. Since there are three means there could be a linear or a quadratic relationship and its significance is tested using F-test as shown in Table 5.4 where we find that there is a significant quadratic trend (F = 210.094, p < 0.001) among the mean

QOL at the three instances. As an option we can also view the plot of means which shows a quadratic line trend.

TABLE 5.4: Tests of within-subjects contrasts

| Source | Model | Type III Sum of Squares | df | Mean Square | F | Sig. |
|---|---|---|---|---|---|---|
| QOL | Linear | 73.205 | 1 | 73.205 | 28.388 | <0.001 |
| | Quadratic | 229.402 | 1 | 229.402 | 210.094 | <0.001 |
| Error | Linear | 255.295 | 99 | 2.579 | | |
| | Quadratic | 108.098 | 99 | 1.092 | | |

5. We can also compare the pairwise differences selecting the estimated marginal means for the factor QOL and use Bonferroni adjustment for comparison. This gives the results shown in Table 5.5.

We observe that all the pairs show significant differences in the mean QOL.

TABLE 5.5: Pairwise Comparisons of QOL values

| (I) QOL | (J) QOL | Mean Difference (I-J) | Std. Error | Sig.[b] | 95% CI for Difference |
|---|---|---|---|---|---|
| QOL_1 | QOL_2 | 1.250* | 0.074 | <0.001 | [1.102, 1.398] |
| | QOL_3 | -1.210* | 0.227 | <0.001 | [-1.661, -0.759] |
| QOL_2 | QOL_1 | -1.250* | 0.074 | <0.001 | [-1.398, -1.102] |
| | QOL_3 | -2.460* | 0.230 | <0.001 | [-2.917, -2.003] |
| QOL_3 | QOL_1 | 1.210* | 0.227 | <0.001 | [0.759, 1.661] |
| | QOL_2 | 2.460* | 0.230 | <0.001 | [2.003, 2.917] |

Based on estimated marginal means.
*. The mean difference is significant at the 0.05 level.
b. Adjustment for multiple comparisons: Bonferroni.

**Remark-1:**

When repeated measurements are made at different time points rather than instances, 'trend analysis' is a relevant study. When the measurements are made at different instances (locations, methods or operators etc.,) the order in which they were defined in the *factor* is important. For instance in the above example we have recorded the levels of QOL as QOL_1, QOL_2 and QOL_3. If the order was changed at the time of defining the *factor* , we get a different shape for the trend. However, with time line data, this problem does not arise.

The next section demonstrates the skills of RM ANOVA using MedCalc.

## 5.3  RM ANOVA Using MedCalc

MedCalc is another software used by clinical researchers for statistical analysis. We can open an SPSS file in this software and run the analysis. The menu sequence for RM ANOVA is *Statistics → ANOVA → Repeated measures Analysis of Variance*. We have to select thee variables in the order QOL_1, QOL_2 and QOL_3. Since there is no factor for grouping the records, we can press OK. The output of MedCalc is shown in Figure 5.3.

| Number of subjects | 100 |
|---|---|

**Sphericity**

| Method | Epsilon |
|---|---|
| Greenhouse-Geisser | 0.581 |
| Huynh-Feldt | 0.583 |

**Test of Within-Subjects Effects**

| Source of variation | | Sum of Squares | DF | Mean Square | F | P |
|---|---|---|---|---|---|---|
| Factor | Sphericity assumed | 302.607 | 2 | 151.303 | 82.44 | <0.001 |
| | Greenhouse-Geisser | 302.607 | 1.162 | 260.527 | 82.44 | <0.001 |
| | Huynh-Feldt | 302.607 | 1.167 | 259.362 | 82.44 | <0.001 |
| Residual | Sphericity assumed | 363.393 | 198 | 1.835 | | |
| | Greenhouse-Geisser | 363.393 | 114.990 | 3.160 | | |
| | Huynh-Feldt | 363.393 | 115.507 | 3.146 | | |

**Trend analysis**

| Trend | t | DF | Significance |
|---|---|---|---|
| Linear | 5.3280 | 99 | P < 0.0001 |
| Quadratic | 14.4946 | 99 | P < 0.0001 |

**Within-subjects factors**

| Factor | Mean | Std. Error | 95% CI | |
|---|---|---|---|---|
| QOL_1 | 60.4200 | 0.3406 | 59.7442 to 61.0958 | |
| QOL_2 | 59.1700 | 0.3405 | 58.4943 to 59.8457 | |
| QOL_3 | 61.6300 | 0.4165 | 60.8036 to 62.4564 | |

**Pairwise comparisons**

| Factors | | | Mean difference | Std. Error | P [a] | 95% CI [a] |
|---|---|---|---|---|---|---|
| QOL_1 | - | QOL_2 | 1.250 | 0.0744 | <0.0001 | 1.069 to 1.431 |
| | - | QOL_3 | -1.210 | 0.227 | <0.0001 | -1.763 to -0.657 |
| QOL_2 | - | QOL_1 | -1.250 | 0.0744 | <0.0001 | -1.431 to -1.069 |
| | - | QOL_3 | -2.460 | 0.230 | <0.0001 | -3.021 to -1.899 |
| QOL_3 | - | QOL_1 | 1.210 | 0.227 | <0.0001 | 0.657 to 1.763 |
| | - | QOL_2 | 2.460 | 0.230 | <0.0001 | 1.899 to 3.021 |

[a] Bonferroni corrected

 Plot means

FIGURE 5.3: MedCalc output of RM ANOVA.

MedCalc also provides a graphical comparison of the variables in several forms by clicking the link *plot means* (in the output). The option window shows 'Multivariable graph options'. We select 3 variables and choose Box-and-whisker plot. After a clicking the 'OK' tab the graph appears as shown in Figure 5.4

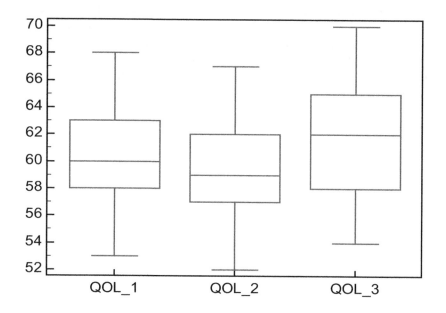

FIGURE 5.4: Box and Whisker plot of QOL.

As an alternative to the box-plot we can also choose another chart like Notched box-and-whisker, Dots (plot all data) etc.

In the following section we discuss the effect of one or more factors on RM ANOVA and illustrate the method.

## 5.4   RM ANOVA with One Grouping Factor

Suppose we wish to compare RM data when there is a grouping factor like the dose of a treatment. Then the different levels of the RM data have to be compared together across various groups. In longitudinal follow-up studies, the objective would be to compare a response in two dimensions viz., at different points of time during the period and among different interventions (treatments).

Consider the following illustration.

**Illustration 5.2** The production of laccase activity (n Kat/ml) in *Cochliobolus Hawaiiensis* on carbon source at different media viz., Glucose, Maltose, Xylose, Galactose and Lactose. The incubation time (inc_time) is measured on different days viz., $2^{nd}$, $4^{th}$, $6^{th}$, and $8^{th}$. The production is observed taking 3 replicates for each medium. A portion of the data is shown in Table 5.6.

TABLE 5.6: Production of laccase activity in a carbon source

| S.No | Media | Incubation time (in days) | | | |
|------|-------|------|------|------|------|
| | | $2^{nd}$ | $4^{th}$ | $6^{th}$ | $8^{th}$ |
| 1 | 1 | 0.24 | 0.32 | 0.64 | 0.50 |
| 2 | 1 | 0.20 | 0.30 | 0.68 | 0.48 |
| 3 | 1 | 0.22 | 0.30 | 0.62 | 0.46 |
| 4 | 2 | 0.26 | 0.38 | 0.50 | 0.42 |
| 5 | 2 | 0.24 | 0.32 | 0.52 | 0.44 |
| 6 | 2 | 0.16 | 0.34 | 0.56 | 0.40 |
| 7 | 3 | 0.30 | 0.86 | 0.98 | 0.45 |
| 8 | 3 | 0.32 | 0.88 | 0.96 | 0.42 |
| 9 | 3 | 0.34 | 0.84 | 0.94 | 0.46 |
| 10 | 4 | 0.18 | 0.20 | 0.45 | 0.30 |
| 11 | 4 | 0.18 | 0.22 | 0.44 | 0.32 |
| 12 | 4 | 0.20 | 0.20 | 0.43 | 0.30 |
| 13 | 5 | 0.16 | 0.24 | 0.30 | 0.20 |
| 14 | 5 | 0.18 | 0.20 | 0.28 | 0.24 |
| 15 | 5 | 0.20 | 0.22 | 0.26 | 0.22 |

(Data Courtesy: Dr. D.V.R. Sai Gopal, Professor,

Department of Virology, Sri Venkateswara University, Tirupati.)

We wish to test whether the production of laccase activity has any effect on the media during the period of study.

**Analysis:**

Let us create a file in SPSS for the above data. Let us define the new RM factor and rename it as inc_time. Then select 'media' as the between subjects factor as shown in Figure 5.5.

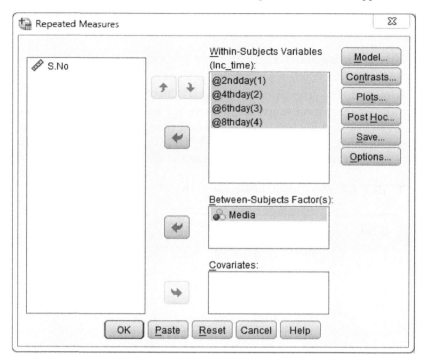

FIGURE 5.5: Defining a Repeated Measures Factor.

In the options tab, let us select estimated marginal means for a) media, b) inc_time and c) media versus inc_time (interaction).

The summary of the SPSS output is presented below.

1. Each medium has exactly 3 measurements at each observation point.

2. The sphericity condition is violated since Mauchly's W = 0.193 and p = 0.014. So the $\epsilon$ based correction is Greenhouse-Geiseer = 0.601.

3. The F-test for the inc_time is significant with F = 483.123 and p < 0.001. Hence production of laccase activity during the study period is statistically significant.

4. There is also a significant interaction (joint effect) of inc_time and media, on the production (F = 78.297, p < 0.001).

5. The production during the $2^{nd}$ to $8^{th}$ day shows all trends (linear, quadratic and cubic) as significant (F = 476.264, p < 0.001). But, we use the quadratic trend as a model to explain the mean changes in production during the study period.

6. The interaction between inc_time and media also shows a quadratic

trend (F = 114.911, p < 0.001) on the production. This indicates that the production of laccase increases with both inc_time and change of medium.

7. Comparison by media levels: SPSS produces a detailed table of differences in the mean values due to the media effect. By default all possible pairs are compared as shown in Table 5.7. We can however report only those that are significant.

TABLE 5.7: Pairwise comparisons of laccase activity due to media effect

| (I) Treatment | (J) Treatment | Mean Difference (I-J) | Std. Error | Sig.[b] | 95% CI for Difference |
|---|---|---|---|---|---|
| Glucose | Maltose | 0.035* | 0.007 | 0.000 | [0.02, 0.05] |
| | Xylose | -0.232* | 0.007 | 0.000 | [-0.247, -0.218] |
| | Galactose | 0.128* | 0.007 | 0.000 | [0.113, 0.143] |
| | Lactose | 0.188* | 0.007 | 0.000 | [0.173, 0.203] |
| Maltose | Glucose | -0.035* | 0.007 | 0.000 | [-0.05, -0.02] |
| | Xylose | -0.267* | 0.007 | 0.000 | [-0.282, -0.253] |
| | Galactose | 0.093* | 0.007 | 0.000 | [0.078, 0.108] |
| | Lactose | 0.153* | 0.007 | 0.000 | [0.138, 0.168] |
| Xylose | Glucose | 0.232* | 0.007 | 0.000 | [0.218, 0.247] |
| | Maltose | 0.267* | 0.007 | 0.000 | [0.253, 0.282] |
| | Galactose | 0.361* | 0.007 | 0.000 | [0.346, 0.376] |
| | Lactose | 0.421* | 0.007 | 0.000 | [0.406, 0.436] |
| Galactose | Glucose | -0.128* | 0.007 | 0.000 | [-0.143, -0.113] |
| | Maltose | -0.093* | 0.007 | 0.000 | [-0.108, -0.078] |
| | Xylose | -0.361* | 0.007 | 0.000 | [-0.376, -0.346] |
| | Lactose | 0.060* | 0.007 | 0.000 | [0.045, 0.075] |
| Lactose | Glucose | -0.188* | 0.007 | 0.000 | [-0.203, -0.173] |
| | Maltose | -0.153* | 0.007 | 0.000 | [-0.168, -0.138] |
| | Xylose | -0.421* | 0.007 | 0.000 | [-0.436, -0.406] |
| | Galactose | -0.060* | 0.007 | 0.000 | [-0.075, -0.045] |

Based on estimated marginal means.
*. The mean difference is significant at the 0.05 level.
b. Adjustment for multiple comparisons: Least Significant Difference (equivalent to no adjustments).

We find all differences are significant.

8. Comparison by inc_time: Table 5.8 shows the differences in the mean values due to inc_time. We note that in the definition of repeated measures factor, SPSS has given the codes as 1= 2nd day, 2 = 4th day, 3 = 6th day and 4 = 8th day. The codes in Table 5.8 are edited accordingly for better understanding. We find all differences are significant except that of 4th day and 8th day.

TABLE 5.8: Pairwise comparisons of laccase activity due to incubation time

| (I) Inc_time | (J) Inc_time | Mean Difference (I-J) | Std. Error | Sig.[b] | 95% CI for Difference |
|---|---|---|---|---|---|
| 2nd day | 4th day | -0.163* | 0.009 | 0.000 | [-0.182, -0.144] |
|  | 6th day | -0.345* | 0.013 | 0.000 | [-0.374, -0.316] |
|  | 8th day | -0.149* | 0.007 | 0.000 | [-0.164, -0.134] |
| 4th day | 2nd day | 0.163* | 0.009 | 0.000 | [0.144, 0.182] |
|  | 6th day | -0.183* | 0.008 | 0.000 | [-0.201, -0.164] |
|  | 8th day | 0.014 | 0.008 | 0.120 | [-0.004, 0.032] |
| 6th day | 2nd day | 0.345* | 0.013 | 0.000 | [0.316, 0.374] |
|  | 4th day | 0.183* | 0.008 | 0.000 | [0.164, 0.201] |
|  | 8th day | 0.197* | 0.009 | 0.000 | [0.178, 0.216] |
| 8th day | 2nd day | 0.149* | 0.007 | 0.000 | [0.134, 0.164] |
|  | 4th day | -0.014 | 0.008 | 0.120 | [-0.032, 0.004] |
|  | 6th day | -0.197* | 0.009 | 0.000 | [-0.216, -0.178] |

Based on estimated marginal means.

*. The mean difference is significant at the 0.05 level.

b. Adjustment for multiple comparisons: Least Significant Difference (equivalent to no adjustments).

9. The interaction between inc_time and media: Table 5.9 shows a two-way table of means obtained at various combinations of days and the media type.

TABLE 5.9: Mean laccase activity by medium and incubation time

| Media | Inc_time | Mean | Std. Error |
|---|---|---|---|
| Glucose | 2nd day | 0.22 | 0.017 |
| | 4th day | 0.31 | 0.012 |
| | 6th day | 0.65 | 0.014 |
| | 8th day | 0.48 | 0.011 |
| Maltose | 2nd day | 0.22 | 0.017 |
| | 4th day | 0.35 | 0.012 |
| | 6th day | 0.53 | 0.014 |
| | 8th day | 0.42 | 0.011 |
| Xylose | 2nd day | 0.32 | 0.017 |
| | 4th day | 0.86 | 0.012 |
| | 6th day | 0.96 | 0.014 |
| | 8th day | 0.44 | 0.011 |
| Galactose | 2nd day | 0.19 | 0.017 |
| | 4th day | 0.21 | 0.012 |
| | 6th day | 0.44 | 0.014 |
| | 8th day | 0.31 | 0.011 |
| Lactose | 2nd day | 0.18 | 0.017 |
| | 4th day | 0.22 | 0.012 |
| | 6th day | 0.28 | 0.014 |
| | 8th day | 0.22 | 0.011 |

The actual output given by SPSS is edited and reproduced for quick understanding.

10. We find from Table 5.9 that the highest production is recorded at the combination of Xylose at the end of the $6^{th}$ day.

11. The graph of inc_time is also shown in Figure 5.6 created in MS-Excel. Though SPSS creates a graph for the mean profiles, we preferred MS-Excel.

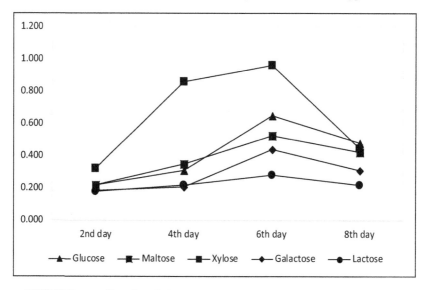

FIGURE 5.6: Graphical display of mean profiles of laccase activity.

12. The mean production due to the media averaged over the days is shown below.

Media

| Production of Laccase | Glucose | Maltose | Xylose | Galactose | Lactose |
|---|---|---|---|---|---|
| Mean | 0.41 | 0.38 | 0.65 | 0.29 | 0.23 |
| 95% CI | [0.403, 0.424] | [0.368, 0.389] | [0.635, 0.656] | [0.274, 0.296] | [0.214, 0.236] |

13. The mean values of production due to days averaged over all media types are shown below.

| Production of Laccase | $2^{nd}$ | $4^{th}$ | $6^{th}$ | $8^{th}$ |
|---|---|---|---|---|
| Mean | 0.23 | 0.39 | 0.57 | 0.37 |
| 95% CI | [0.209, 0.242] | [0.376, 0.4] | [0.557, 0.584] | [0.363, 0.385] |

Thus RM analysis can be used to assess the statistical significance of the influence of factors on the outcome that has been measured at different instances on the same subject of study. We can also include one or more continuous variables as *cofactors* or *covariates* and examine their influence on the estimated changes over time, on the outcome variable.

In the following section we introduce a technique called *profile analysis* which is used to compare the profiles of several variables which are measured in the same units.

---

## 5.5 Profile Analysis

An interesting application of RM analysis is the *profile analysis*. A graphical representation of the means of RM data is known as a *profile plot*.

Let $X_1$, $X_2$, ..., $X_k$ denote k-variables measured on the *same scale*. We say these variables are *commensurate*. For instance in the QOL data all the variables are commensurate.

If $\mu_1, \mu_2, \ldots, \mu_k$ denote the mean values of the k-commensurate variables, then the plot of means against the variables is called a *profile plot*. Similarly the *growth of a culture* at $2^{nd}$, $4^{th}$, $6^{th}$, and $8^{th}$ days of observation is an example of profile analysis.

In general profile analysis deals with statistical analysis of profiles.

Three different types of profile analysis are possible to answer the following questions as hypotheses.

1. The one sample profile analysis (without any prognostic factor) where the question is $H_0$: *Is the profile flat (or level)?*

2. Two-sample profile analysis with a factor at two levels. The questions to be answered are:

   a) $H_{01}$: *Are the profiles flat?*

   b) $H_{02}$: *Are the profiles parallel?*

3. k-sample profile analysis with a factor at more than two levels. The questions to be answered are:

   a) $H_{01}$: *Are the profiles flat?*

   b) $H_{02}$: *Are the profiles parallel?*

   c) $H_{03}$: *Are the profiles coincident (having equal levels)?*

The analysis goes with the methods used in the MANOVA. Here are simple tips to handle this analysis with SPSS, for each context.

**One sample profile analysis**

Let us consider the QOL data used in Illustration 5.1. The *one-sample profile plot* can be created as a part of RM ANOVA by selecting the *plot* option.

The profile plot for the QOL data is shown in Figure 5.7. Since the QOL panel has three different variables but observed on the same patient, measured on the same scale, they are commensurate.

We have renamed the within-subjects factor as QOL with 3 levels and no factor is selected. Using the *plot* option and taking QOL to the horizontal axis, we click on the *add* tab. The profile plot is created based on the estimated marginal mean values at the three QOL variables.

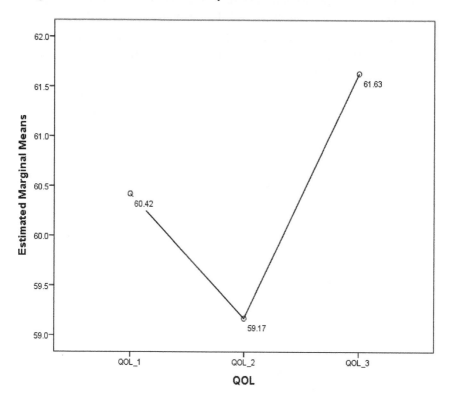

FIGURE 5.7: Profile plot of QOL.

We observe that the QOL profile is not flat. If the connecting line was *nearly horizontal*, we may consider it as flat but it is not so here. Had the mean values been more or less equal, the plot would have been nearly horizontal.

A visually sloping profile is not a flat one and indicates a state of unequal mean values for the variables in the profile. The statistical significance of the deviation from flatness is an important criterion.

We then wish to test the hypothesis $H_0$: $\mu_1 = \mu_2 = \ldots = \mu_k$ against $H_1$: $\mu_i \neq \mu_j$ for some $i \neq j$. Since all the k-variables are correlated to each, we have to use the RM ANOVA which produces the following results.

1. The sphericity test shows Mauchly's $W = 0.278$ with $p < 0.001$ and hence the F- test is to be taken after adjusting for degrees of freedom.

2. The Greenhouse-Geisser method shows $F = 82.440$ and $p < 0.001$ and hence the hypothesis of a flat profile cannot be accepted.

3. The test of linear trend also shows significance but the quadratic is more significant. (The sample plot also looks convex!)

### Profile analysis with a prognostic factor at k-levels:

Let us continue analyzing the QOL data of the previous section. Suppose there is a factor like *gender* (with $k = 2$ levels) and we wish to compare the QOL profiles of male and female patients.

The following steps help in getting the profile plot using SPSS.

1. Go to 'Repeated Measures window' as shown in Figure 5.5.

2. Send gender to the 'Between-Subjects Factor(s)'.

3. Click on the 'Plots' tab.

4. Select the factors as shown in Figure 5.8.

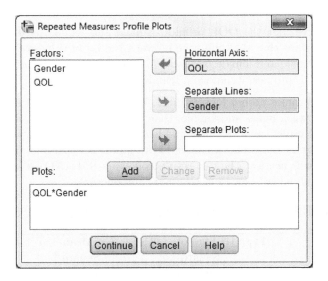

FIGURE 5.8: SPSS profile plot options.

5. 'Continue' → 'OK'.

This gives a separate profile plot of mean QOL for male and female subjects as shown in Table 5.9.

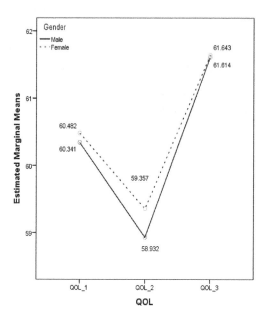

FIGURE 5.9: Profile plot of QOL by gender.

We observe that the QOL profiles of male and female patients are not parallel. It means that the difference between the mean values of QOL is not the same for male and female patients. We wish to test the hypothesis where the mean QOL profiles of male and female patients are parallel.

The two profiles do not look parallel. The means of QOL_1 and QOL_3 are closer between males and females than the mean of QOL_2. However, the violation of parallelism of the male and female profiles should be judged on its statistical significance. The within-subjects factor is QOL and the between-subject factor is *gender*. The test results following RM ANOVA are shown in Table 5.10 (edited to show only required components).

TABLE 5.10: Tests of within-subjects effects of QOL and gender

| Source | Type III Sum of Squares | df | Mean Square | F | Sig. |
|---|---|---|---|---|---|
| QOL | 304.055 | 1.156 | 262.943 | 82.464 | <0.001 |
| QOL * Gender | 2.055 | 1.156 | 1.777 | 0.557 | 0.481 |

*Method: Greenhouse-Geisser.*

We observe the following.

1. The QOL profile is not flat. It means the means differ significantly (F = 82.464, p < 0.001).

2. The interaction between QOL and *gender* is not significant (F = 0.557, p = 0.481). The Greenhouse-Geisser method was used, since Mauchly's test showed lack of sphericity. It means that the hypothesis of parallel profiles cannot be accepted.

3. Hence the QOL profiles of male and female patients cannot be considered as parallel.

**Coincident profiles:**

If the profile lines coincide with each, we say they are coincident. This happens only when the mean values do not differ for each level of the factor.

For instance when the QOL_1, QOL_2 and QOL_3 happen to be the same for both male and female patients, the profiles become coincident. In other words, by examining coincidence, we can know whether one group (male or female) scores higher than another across the 3 measures of QOL.

SPSS provides this as a test of contrast (difference of means between male and female groups) as shown in Table 5.11 which is reported as a *univariate test*.

Each pairwise comparison is called a *contrast* and in this case the contrast refers to the difference between the overall mean QOL of male and female patients.

This can be done by finding out the grand mean of QOL at the 3 measures for both male and female patients and comparing the means. This is a simple univariate comparison which can be done by the F-test.

TABLE 5.11: Univariate tests of QOL profile

Measure: QOL

|  | Sum of Squares | df | Mean Square | F | Sig. |
|---|---|---|---|---|---|
| Contrast | 0.972 | 1 | 0.972 | 0.078 | 0.780 |
| Error | 1215.824 | 98 | 12.406 |  |  |

In this case we get F = 0.078 with p = 0.780. So we cannot reject the hypothesis that the profiles are coincident (having equal levels). It means we accept that the profiles are in general coincident.

When the factor has k ( > 2) groups we get (k-1) contrasts to be tested for significance.

For instance in the case of the production of laccase activity data discussed in Illustration 5.2, the profile factor (medium) has 5 levels. Hence the five lines in Figure 5.10 are an example of a k-level profile plot.

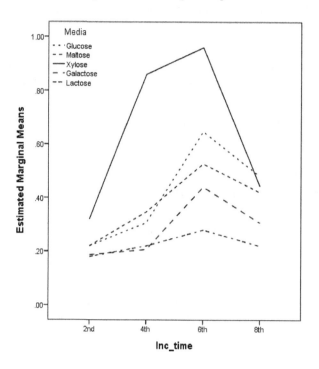

FIGURE 5.10: Profile plot with k-levels.

For questions relating to flatness, parallelism of the profile can be answered as done in the case of QOL data.

We may observe from Figure 5.9 that the profile has a non-linear trend over the incubation time. A profile plot where a single variable is measured at different time intervals is often called a *growth curve*.

We end this chapter with the observation that RM ANOVA alone can give a correct picture of changes in longitudinal studies where observations are made on the same subject at various instances and need comparison.

# Summary

RM ANOVA is the appropriate tool for comparing the means of several correlated variables simultaneously. The statistical principle is based on the comparison of multivariate mean vectors using a) Hotelling's test or other similar tests and b) Mauchly's test of sphericity. Once the RM ANOVA shows the significance of the effect of the RM factor, we can proceed with univariate comparisons, using conventional ANOVA followed by multiple comparison tests. An interesting application of RM ANOVA is the profile analysis, used to compare the means of repeated measures of variables observed on the same scale. One can use software like SPSS to perform this analysis.

# Do it yourself (Exercises)

**5.1** When $X_1, X_2, X_3$ are three correlated variables, the differences $d_1 = X_1 - X_2, d_2 = X_1 - X_3$ and $d_3 = X_2 - X_3$ often tend to be uncorrelated. Reconsider the data used in Illustration 5.1 and carry out RM ANOVA on $d_1, d_2, d_3$ and observe the change in the results.

**5.2** The estimated marginal means and standard errors provide valuable information on the effects under study. Obtain the estimated marginal means of QOL for the data used in Illustration 5.1 and plot them using MS-Excel. Also include the standard errors around the means.

**5.3** The following data refers to the HDL values measured on 30 patients on 4 different occasions under two groups. The variables are HDL_1, HDL_3, HDL_5 and HDL_6 and the two groups represent male and female subjects. Perform RM ANOVA and determine the nature of the profile.

| Male | | | | Female | | | |
|---|---|---|---|---|---|---|---|
| HDL_1 | HDL_3 | HDL_5 | HDL_6 | HDL_1 | HDL_3 | HDL_5 | HDL_6 |
| 46.00 | 46.00 | 48.60 | 48.99 | 54.32 | 54.44 | 58.90 | 59.99 |
| 47.78 | 47.45 | 50.09 | 50.00 | 47.21 | 46.99 | 50.12 | 53.00 |
| 49.57 | 50.10 | 51.45 | 51.88 | 53.23 | 54.00 | 57.00 | 57.99 |
| 47.00 | 46.00 | 48.00 | 49.00 | 48.98 | 48.00 | 53.00 | 53.00 |
| 47.32 | 45.40 | 49.00 | 49.00 | 52.23 | 53.00 | 56.00 | 56.00 |
| 43.34 | 42.00 | 43.00 | 43.00 | 49.00 | 49.00 | 52.00 | 52.00 |
| 49.21 | 48.00 | 50.00 | 50.00 | 47.10 | 47.00 | 50.00 | 50.69 |
| 42.66 | 43.00 | 47.00 | 47.00 | 46.76 | 46.00 | 49.00 | 49.00 |
| 44.43 | 43.00 | 46.00 | 47.00 | 50.50 | 51.00 | 53.00 | 54.00 |
| 46.67 | 46.00 | 48.00 | 49.00 | 48.34 | 48.00 | 51.00 | 52.00 |
| 48.00 | 47.00 | 48.00 | 49.00 | 49.36 | 49.00 | 52.00 | 53.00 |
| 43.32 | 43.00 | 47.00 | 48.00 | 47.71 | 47.00 | 49.00 | 50.00 |
| 44.21 | 43.00 | 46.00 | 48.00 | | | | |
| 41.30 | 42.00 | 46.00 | 47.00 | | | | |
| 44.43 | 44.00 | 47.00 | 48.00 | | | | |
| 49.10 | 49.00 | 50.00 | 51.00 | | | | |

(Data Courtesy: Dr. Kanala Kodanda Reddy, Department of Anthropology, Sri Venkateswara University, Tirupati.)

## Suggested Reading

1. Alvin C.Rencher, William F. Christensen 2012. *Methods of Multivariate Analysis*. 3$^{rd}$ ed. Brigham Young University: John Wiley & Sons.

2. Huynh, H., and Feldt, L.S. 1976. Estimation of the Box correction for degrees of freedom from sample data in randomised block and split-plot designs. *Journal of Educational Statistics*, 1, 69–82.

3. Johnson, R.A., & Wichern, D.W. 2014. *Applied multivariate statistical analysis*, 6$^{th}$ ed. Pearson New International Edition.

4. Kleinbaum, D.G., Kupper, L.L., Muller, K.E., and Nizam, A. 1998. *Applied Regression Analysis and Other Multivariable Methods*, 3$^{rd}$ ed. Belmont, CA, US: Thomson Brooks/Cole Publishing Co.

5. Himavanth Reddy Kambalachenu, Thandlam Muneeswara Reddy, Sirpurkar Dattatreya Rao, Kambalachenu Dorababu, Kanala Kodanda Reddy, K.V.S. Sarma. 2018. A Randomized, Double Blind, Placebo Controlled, Crossover Study to Assess the Safety and Beneficial Effects of Cassia Tora Supplementation in Healthy Adults. *Reviews on Recent Clinical Trials*, 13(1). DOI: 10.2174/1574887112666171120094539.

# Chapter 6

# Multiple Linear Regression Analysis

| | | |
|---|---|---|
| 6.1 | The Concept of Regression | 115 |
| 6.2 | Multiple Linear Regression | 118 |
| 6.3 | Selection of Appropriate Variables into the Model | 120 |
| 6.4 | Predicted Values from the Model | 126 |
| 6.5 | Quality of Residuals | 128 |
| 6.6 | Regression Model with Selected Records | 129 |
| | Summary | 130 |
| | Do it yourself (Exercises) | 131 |
| | Suggested Reading | 134 |

Statistics: A gateway to knowledge.

C.R. Rao (1920 – )

## 6.1 The Concept of Regression

The word 'regression' refers to a statistical procedure used to fit a mathematical model relating the dependent variable (outcome of the study) with one or more independent (predictors) variables. It is assumed that each independent variable has a non-zero correlation with the dependent variable and all the independent variables are relevant to the context of study.

We know that correlation coefficient explains the strength of the linear relationship between two variables X and Y, but it will not explain the cause and effect relationship. In fact, Y may affect X or the other way around, but the correlation coefficient remains the same.

115

In the context of study, one should state which variable denotes the 'cause' and which denotes the 'effect'. For this reason, a study of simple correlation coefficient alone is not sufficient to understand the true relationship between X and Y.

On the contrary, regression analysis deals with a *cause* and *effect* study between Y and X by building a mathematical model. With only one dependent and another independent variable, the regression is known as *simple linear regression*. If more than one independent variable influences the dependent variable, we refer to the model as *multiple linear regression*.

We use the word linear to mean that Y and X bear a constant proportional relationship. For instance, in a restaurant if the cost (X) of a soft drink is \$3 per can then the total cost (Y) for 20 cans will be \$60 and for 50 cans it will be \$150. It means Y increases at a constant rate of 3 per one unit of X (can). When the number of cans purchased is zero (X = 0), we get Y = 0. Sometimes when there is a price discount for large sales, the linearity will not hold well.

Statistical modeling of the above situation makes use of a linear equation $Y = \beta_0 + \beta_1 X$, where $\beta_0$ is a constant representing the baseline value of Y and $\beta_1$ denotes the rate of change in Y due to a unit change in X. In the soft drink example, $\beta_0$ is something like a fixed 'entry fee' into the restaurant, say \$10 and $\beta_1$ will be \$3. In this case even if no can is purchased we get a bill of Y = \$10!

In general, there will be several variables influencing the values of Y but not all of them will have the same effect on Y. Further, there may be some uncontrollable factors which may show variation in the values of Y. The presence of such factors is called *random error* or *noise*. This noise term is always present in the model, denoted by $\varepsilon$ (epsilon) and a simple linear regression is expressed as the sum of these effects denoted by

$$Y = \beta_0 + \beta_1 X + \varepsilon \tag{6.1}$$

It is assumed that $\varepsilon$ is a random variable following normal distribution with mean 0 and variance $\sigma_\varepsilon^2$. It means that the uncontrollable errors *settle to zero on average* but will have some variance. While we cannot eliminate this error component, we can design a method of estimating $\beta_0$ and $\beta_1$ in such a way that this error variance is minimized.

Once we estimate $\beta_0$ and $\beta_1$, we can predict the 'average value of Y' at a given value of X by substituting X values in Equation 6.1.

The values of $\beta_0$ and $\beta_1$ are usually unknown but can be estimated from sample data by using a method called *method of least squares*. The idea is to find such values of $\beta_0$ and $\beta_1$ which would minimize the squared error between the actual Y and the predicted Y. $\beta_1$ is called the *regression coefficient* and $\beta_0$ is called the *intercept*.

Consider the following illustration.

**Illustration 6.1** The Body Mass Index (BMI) is known to increase with an increase in Waist Circumference (WC) in cms. Values of WC and BMI obtained in a study from 20 persons are shown in Table 6.1. We wish to explain the relationship visually and express it numerically

TABLE 6.1: Paired values of WC and BMI

| S.No | 1 | 2 | 3 | 4 | 5 | 6 | 7 | 8 | 9 | 10 |
|------|------|------|------|------|------|------|------|------|------|------|
| WC | 99 | 88 | 78 | 68 | 92 | 104 | 95 | 92 | 103 | 78 |
| BMI | 27.4 | 29.3 | 23.5 | 18.4 | 26.5 | 31.2 | 25.4 | 27.3 | 30.0 | 22.0 |

| S.No | 11 | 12 | 13 | 14 | 15 | 16 | 17 | 18 | 19 | 20 |
|------|------|------|------|------|------|------|------|------|------|------|
| WC | 97 | 76 | 105 | 109 | 104 | 110 | 123 | 115 | 87 | 107 |
| BMI | 30.5 | 22.6 | 34.5 | 36.5 | 32.6 | 39.6 | 39.7 | 34.0 | 30.5 | 33.2 |

**Analysis:**

The relationship between BMI and WC can be visually understood with the help of a *scatter diagram* as shown in Figure 6.1 by using MS-Excel charts. We find that the spread of values shows an upward (increasing) trend indicating that as WC increases, the BMI also increases.

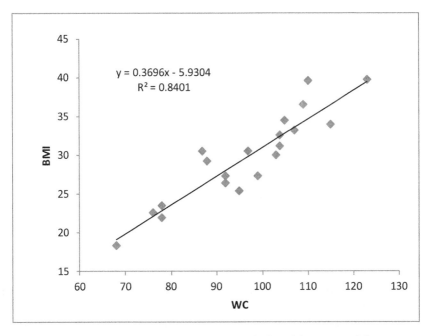

FIGURE 6.1: Scatter diagram with fitted linear model.

The best line that passes through as many points as possible is the desired line. This can be obtained from the scatter diagram of MS-Excel with the following steps.

1. Right click on any data point.

2. Choose *Add trend line* option.

3. Select the options *show equation on the chart*, and *show R-square*.

4. Do not select the option *Set intercept zero*.

This gives the line Y = 0.3696*X-5.9304. We may also write this as Y = -5.9304+0.3696*X so that $\beta_0$ = -5.9304 and $\beta_1$ = 0.3696.

The model also shows a value $R^2$ = 0.8401. It means that about 84% of the BMI pattern can be predicted with the knowledge of WC. The value $R^2$ is a simple measure of adequacy of the linear model that has been fitted to the data. We will see more details about $R^2$ in the following sections. At this point we only observe that a higher value of $R^2$ usually represents a better model (but need not always be true).

**Why fit a model?**

The purpose of fitting a regression model (linear or a curve) is to characterize the behavior of Y in terms of X and to estimate the average Y value for a new case where X is known (but not Y), provided that the X value is not an outlier (abnormal).

In the following section, we extend the logic of simple linear regression to the case where we have two or more independent variables.

---

## 6.2    Multiple Linear Regression

A multiple linear regression model will be of the form

$$Y = \beta_0 + \beta_1 X_1 + \beta_2 X_2 + \ldots + \beta_k X_k + \varepsilon \tag{6.2}$$

where $\beta_1, \beta_2, \ldots, \beta_k$ are constants called *partial regression coefficients* and $\beta_0$ is the intercept. Y is the *dependent variable* and $X_1$, $X_2$, ... , $X_k$ are the *independent variables* or *explanatory variables* or *regressors* or *predictor variables*. The term $\varepsilon$ is the *error component* and $\varepsilon$ represents the error in estimating the value of Y from the $i^{th}$ sample of data.

The individual data records can be expressed as follows.

$$\left.\begin{aligned}
Y_1 &= \beta_0 + \beta_1 X_{11} + \beta_2 X_{12} + \ldots + \beta_p X_{1k} + \varepsilon_1 \\
Y_2 &= \beta_0 + \beta_1 X_{21} + \beta_2 X_{22} + \ldots + \beta_p X_{2k} + \varepsilon_2 \\
&\ldots \\
Y_n &= \beta_0 + \beta_1 X_{n1} + \beta_2 X_{n2} + \ldots + \beta_p X_{nk} + \varepsilon_n
\end{aligned}\right\} \tag{6.3}$$

For the $i^{th}$ record, $Y_i$ is the observed outcome, $X_{ij}$ represents the value on the $j^{th}$ predictor and $\varepsilon_i$ denotes the error in predicting $Y_i$.

It is also assumed that $\varepsilon$ has mean zero and variance $\sigma^2$ and the covariance between any pair of error terms is zero. It means that the error terms are uncorrelated to each other.

The regression coefficient $\beta_i$ represents the marginal change in Y corresponding to a unit change in $X_i$ ; i=1,2,...,k. The values of $\beta's$ are estimated from sample data by using the method of *least squares*.

If $\mathbf{X}$ denotes the design matrix of the X variables, containing the data on the predictor variables, then the least squares estimate of the regression coefficients is given by the matrix equation $\widehat{\beta} = (\mathbf{X'X})^{-1}(\mathbf{X'Y})$ where $\widehat{\beta}$ denotes the vector of (k+1) regression coefficients.

Here are some observations regarding the linear regression model.

1. With the estimated regression coefficients, denoted by $\hat{\beta}_j$, the fitted model is stated as $\widehat{Y} = \widehat{\beta}_0 + \widehat{\beta}_1 X_1 + \widehat{\beta}_2 X_2 + \ldots + \widehat{\beta}_k X_k$.

2. The statistical significance of the fitted regression coefficients $\widehat{\beta}_j$ is tested by t-test. The null hypothesis is stated about the unknown population regression coefficients. Hence, $H_0 : \beta_j = 0$ for each j.

3. Given the values of $X_1, X_2, \ldots, X_k$ for the $i^{th}$ sample, it is possible to estimate the value of Y, denoted by $\widehat{Y}_i$ with the help of the above model. This estimate is the average value of Y corresponding to the given X values and will have some *standard error*. However, $\widehat{Y}_i$ cannot be considered as the exact value of Y predicted by the model. It only says that given the information on the X-variable, the model predicts the average value of Y. For this reason, such a prediction carries some *standard error*.

4. The *goodness of fit* of the model is measured in terms of the *squared multiple correlation coefficient* given by $R^2 = \dfrac{\sum (Y_i - \widehat{Y}_i)^2}{\sum (Y_i - \overline{Y})^2}$.

5. The value of $R^2$ lies between 0 and 1.

6. The statistical significance of the multiple correlation coefficient (R) is tested by F-test under the null hypothesis that R = 0 in the population.

7. The model is considered as *adequate* or *a good fit*, if the $R^2$ value is high and close to 1. If $R^2 = 0.86$, it means that all the regressors used in the equation account for about 86% of the variation in Y. It is also possible that when a large number of variables are used as regressors, the $R^2$ value tends to increase, even though some regressors are either unimportant or irrelevant!

8. Another criterion to measure the adequacy of the fitted model is the *adjusted R-square* given by

$$\overline{R}^2 = 1 - \frac{(n-1)}{(n-k)}(1 - R^2)$$

where k is the number of independent variables. The difference is that $\overline{R}^2$ may decrease as k increases, if the decrease in $(n-k)(1-R^2)$ is not compensated for by the loss of one degree of freedom in (n-k). It is suggested that *adjusted $R^2$* is used in place of $R^2$.

The calculations for fitting a multiple linear regression model are quite complicated and there are many software packages to fit multiple linear regression and SPSS is one such. The R software has more options for an in-depth study of these models.

In the next section we discuss a method for selecting only a few appropriate variables, into the model.

---

## 6.3   Selection of Appropriate Variables into the Model

Sometimes, the researcher may consider several X variables capable of influencing Y and put them into the model. However, all of them may not contribute significantly to explain Y. It is important to select the most promising explanatory variables alone into the model instead of all. So we need a method of selecting appropriate variables into the model and leave those which fail to show significant contribution to the model. This approach is often known as *stepwise regression*.

The SPSS steps for running Regression are given below.

1. Analyze → Regression → Linear.

2. Select the 'Dependent' variable.

3. Choose all the possible 'Independent(s)' variables.

4. Click the tab 'Method' for selection of the appropriate variable into the model. This shows five types of procedures as discussed below.

   a) **Enter:** In this method, all the independent variables will be entered into the equation simultaneously.

   b) **Stepwise:** With this option, SPSS selects the variables one after another in a sequential way. A variable which is already not included in the model will be included, provided that it has the smallest p-value of the F-test. The method also removes one variable from among those already included in the equation, for which, the p-value and F-test is very large. The method stops when no variable is eligible for entry or removal.

   c) **Remove:** In this method, all the independent variables will be removed from the model. This cannot be used as an option while building the model.

   d) **Backward:** In this method all the variables will be first entered in the model and specific variables are removed sequentially. The removal is based on what is called *partial correlation coefficient*. A variable which has the smallest partial correlation coefficient with Y will be first removed. In the next step the same criterion is adopted to remove another variable from those available. The method stops when there is no variable that satisfies the removal criterion.

   e) **Forward:** This is a stepwise procedure in which a variable having the largest partial correlation coefficient with the dependent variable is selected for entry into the model, provided that it satisfies the F-test criterion for entry. Then, another variable will be selected in the same way and the procedure is repeated until there is no variable eligible for entry.

5. From the 'Statistics' tab we can select a) Estimate, b) Model fit and c) R squared change. The other options are checked only when required. When the nature of the residuals is required we may check the options under the group 'Residuals'.

6. The entry/removal of variables into the regression model is based on the F-test for the $R^2$ value. The value of the F-ratio or the p-value of the F-test can be used for this operation. By default, the p-value for entry of a variable is taken as 0.05 and for removal it is taken as 0.10. The user can modify these values before running the model.

7. Click 'OK'.

Consider the following illustration.

**Illustration 6.2** Let us reconsider the Mets data used in Illustration 2.1.

The researcher wants to develop a model relating LVMPI with various factors like age, BMI, Waist Circumference (WC), TGL, HDL, Gender (Female = 0, Male = 1), DM (No = 0, Yes = 1, Pre-diabetic = 2 ), Metabolic Syndrome (MetS) (No = 0, Yes = 1) etc.

The objective is to:

a) Estimate the effect of factors on LVMPI;

b) Identify the most influencing factors and;

c) Predict the likely LVMPI of a new patient for whom the data on the factors is available (but not LVMPI).

**Analysis:**

The following are the steps to run the regression in SPSS.

1. Open the data file. Choose Analyse → Regression → Linear

2. In the resulting dialogue box select the dependent and independent variables as shown in Figure 6.2.

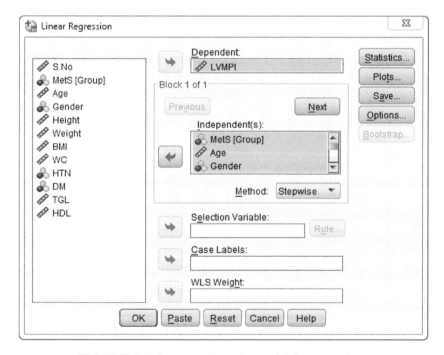

FIGURE 6.2: Input options for multiple regression.

3. Choose the method as *stepwise*.

4. Click on the 'Statistics' button and choose *estimates* (of the regression coefficients), *model fit* and *R-squared change*. Press the Continue button.

5. Click on the 'Options' button and choose the default options, unless we wish to change any of them.

6. A constant is included in the model by default and we can retain this option.

7. The other options relate to 'Saving' of the predicted values and plotting of the residuals (errors). We will discuss these aspects later.

8. Press OK and the results appear in a separate output file.

From the regression analysis several outputs are provided and one needs a detailed interpretation. Here are the salient features of the analysis.

**Model summary:**

The stepwise procedure has been completed in 4 steps and the stepwise improvement of the model is shown in the following Table 6.2.

TABLE 6.2: Model summary of regression

| Model | R | R Square | Adjusted R Square* | Change Statistics | | | | |
|-------|---|----------|--------------------|-------------------|--------|-----|-----|--------------|
|       |   |          |                    | R Square Change | F Change | df1 | df2 | Sig. F Change |
| 1 | 0.676[a] | 0.457 | 0.450 (0.0833) | 0.457 | 65.570 | 1 | 78 | 0.000 |
| 2 | 0.707[b] | 0.500 | 0.487 (0.0804) | 0.043 | 6.629 | 1 | 77 | 0.012 |
| 3 | 0.738[c] | 0.545 | 0.527 (0.0773) | 0.045 | 7.467 | 1 | 76 | 0.008 |
| 4 | 0.757[d] | 0.573 | 0.551 (0.0753) | 0.029 | 5.058 | 1 | 75 | 0.027 |

*. Figures in braces indicate std. error of the estimate.
a. Predictors:(Constant), MetS.
b. Predictors:(Constant), MetS, DM.
c. Predictors:(Constant), MetS, DM, WC.
d. Predictors:(Constant), MetS, DM, WC, Age.

**Change in $R^2$:**

We observe that the variables are selected into the model one by one in such a way that there is a significant improvement in the R-square at each step. The change in R-square is found decreasing from step to step. At the end of the 4th step the procedure is terminated because the improvement in $R^2$ is not significant. For the purpose of further interpretation we use the 4th model having $R^2 = 0.757$.

**Model termination:**

The model summary also shows the standard error of the estimate as 0.075.

It is the standard error of LVMPI when estimated by using the model having only the factors MetS, DM, WC and Age. The last column in Table 6.2 shows the p-value of the F-test based on which the stepwise method terminates. In this case we observe that the change from step to step has $p < 0.05$. The algorithm automatically terminates in the next step when this p-value exceeds the limit.

**Significance of the model:**

The statistical significance of the model developed is shown by ANOVA wherein we find that the F-ratio due to regression is 25.192 with p-value < 0.001. It means that the regression model developed from the sample data is statistically significant.

The regression coefficient corresponding to each independent variable along with its standard error, t-value and the p-value will be displayed as shown in Table 6.3. The sign of the regression coefficient denotes the direction in which Y changes in tune with the change in X.

TABLE 6.3: Regression coefficients and their significance in the final model

| Model [a] | | Unstandardized Coefficients | | Standardized Coefficients | t | Sig. |
|---|---|---|---|---|---|---|
| | | B | Std. Error | Beta | | |
| 4[*] | (Constant) | 0.201 | 0.082 | | 2.455 | 0.016 |
| | MetS | 0.171 | 0.022 | 0.742 | 7.878 | 0.000 |
| | DM | -0.053 | 0.016 | -0.292 | -3.314 | 0.001 |
| | WC | 0.002 | 0.001 | 0.257 | 3.041 | 0.003 |
| | Age | 0.002 | 0.001 | 0.176 | 2.249 | 0.027 |

a. Dependent Variable: LVMPI.
*. 4 indicates, model-4 of stepwise regression.

**Regression coefficients:**

The unstandardized coefficient for a variable under the column titled 'B' represents the marginal contribution of that variable to LVMPI. For instance, the B value for MetS shows that when compared to those without MetS, those with MetS will have an additional increase of 0.171 units of LVMPI (keeping other things unchanged). Since this variable is categorical with a yes/no option, the coefficient is interpreted as the change in LVMPI due to a change from *no* to *yes* status.

Similar observation can be made regarding DM where the marginal changes are a decrease since the coefficient is -0.053. With regards to the waist circumference (WC), the coefficient 0.002 is the marginal increase in LVMPI due to one unit (cm) increase in WC.

**Relative importance of variables and beta coefficients:**

The relative importance of the predictor variables in the model is expressed in terms of values called *beta coefficients*. The regression coefficient of a variable in the fitted model, when all variables are in their *standardized z-score form*, is called *beta coefficient*. The larger beta coefficient (absolute value), the more important it is that variable in explaining the response.

Here we find that MetS has the highest relative importance (0.742), followed by DM, WC and age. This is one method of variable selection. This helps as a screening step when there is a large number of variables finding a place in the regression model.

This is needed to ensure that the observed B-value is not an occurrence by chance! The null hypothesis assumes zero value for each regression coefficient and its truth is tested by a t-test. The t-test for each variable indicates a test for the significance of the regression coefficient. We see that all the variables and the constant are statistically significant ($p < 0.05$).

**Excluded variables and collinearity:**

We also get an interesting and important output that indicates the variables that were excluded by the stepwise procedure. Table 6.4 shows the list of excluded variables along with relevant statistical information on these variables.

TABLE 6.4: Variables excluded from the model

| Variable | Beta In | t | Sig. | Partial Correlation | Collinearity Statistics Tolerance |
|---|---|---|---|---|---|
| Gender | 0.056 | 0.716 | 0.476 | 0.083 | 0.929 |
| BMI | -0.060 | -0.548 | 0.585 | -0.064 | 0.478 |
| TGL | 0.107 | 1.325 | 0.189 | 0.152 | 0.865 |
| HDL | -0.055 | -0.67 | 0.505 | -0.078 | 0.856 |

We find that Gender, BMI, TGL and HDL were not included in the final model. The t-test shows no significance of the regression coefficient for all these variables ( $p > 0.05$).

The partial correlation of each variable with LVMPI is also very low and hence these variables *could not find place* in the model.

The collinearity statistics is a parameter that shows a measure called *tolerance* which measures a characteristic called *multicollinearity*, a feature that addresses redundancy among the predictors.

A tolerance of less than 0.20 or 0.10 indicates the presence of multicollinearity. When some of the independent variables are correlated among themselves, we say there is a problem of multicollinearity which distracts from the true model. In the present case all the tolerance levels are above 0.10 and hence there is no multicollinearity.

**Final model:**

The final results can be presented as follows.

1. Independent variables: MetS, DM, WC, Age, BMI, TGL and HDL.

2. Method used: Stepwise Linear Regression.

3. Variables included in the model: MetS, DM, WC and Age.

4. Variables excluded: BMI,TGL, HDL, Gender.

5. Model:

$$\text{LVMPI} = 0.201 + 0.171*\text{MetS} - 0.053*\text{DM} \\ + 0.002*\text{WC} + 0.002*\text{Age} \tag{6.4}$$

We can use this model to estimate the LVMPI of a new patient for whom the above variables are measured.

**Remark-1:**

Suppose we have not used stepwise regression but used 'method = Enter'. Then we would get a different model (try this!). We get $R^2 = 0.590$ which is higher than 0.573 obtained with the stepwise method and one may be tempted to conclude that the *full regression* (with the *enter* method) gives a better fit. But the adjusted $R^2$ in the stepwise regression was 0.551 while it is only 0.544 in the full regression. This is due to the fact that $R^2$ value can increase, when more and more regressor variables are included in the model, though some of them contribute insignificantly. The correct way of running regression is by keeping the most promising variables alone and the stepwise regression will do this.

In the following section we examine the issues of predicated values from the model and understand the nature of residuals.

---

## 6.4   Predicted Values from the Model

Once the model is built, we can predict the response (LVMPI) by using the model obtained in Equation 6.4. For each data case, SPSS gives the predicted response along with 95% confidence interval for the true value.

These values will be saved to the data file by selecting the 'Save' tab and check the option 'Unstandardized' from the 'Predicated Values' group as well as the 'Residuals' group.

Under 'Save' tab there are other *diagnostic statistics* which will be saved to the data file for evaluating the goodness of the fitted model. The advantages of saving all these statistics is beyond the present scope of discussion.

The difference between the actual value and the predicted value for a case is called the *residual* which is also known as the *unstandardized residual*. The residual can also be standardized by subtracting from each residual, the mean and dividing by the standard deviation.

In the present model we have chosen to save a) unstandardized predicted values, b) prediction intervals for individual values, and c) unstandardized residuals.

A portion of the saved values (for the first 10 cases) are shown in Table 6.5.

TABLE 6.5: Prediction of individual LVMPI values from the regression model

| Case ID | Actual | Predicted | Residual | Lower CI | Upper CI |
|---------|--------|-----------|----------|----------|----------|
| 1 | 0.73 | 0.61048 | 0.11952 | 0.45832 | 0.7626 |
| 2 | 0.66 | 0.64600 | 0.01400 | 0.49341 | 0.7986 |
| 3 | 0.60 | 0.63843 | -0.03843 | 0.48614 | 0.7907 |
| 4 | 0.52 | 0.58678 | -0.06678 | 0.43313 | 0.7404 |
| 5 | 0.60 | 0.60104 | -0.00104 | 0.44839 | 0.7537 |
| 6 | 0.75 | 0.61538 | 0.13462 | 0.46333 | 0.7674 |
| 7 | 0.50 | 0.63036 | -0.13036 | 0.47855 | 0.7822 |
| 8 | 0.73 | 0.64535 | 0.08465 | 0.49326 | 0.7974 |
| 9 | 0.61 | 0.65661 | -0.04661 | 0.50232 | 0.8109 |
| 10 | 0.52 | 0.61681 | -0.09681 | 0.46427 | 0.7694 |

We observe that the actual and predicted LVMPI values do have some difference as shown in the column *residual*. The confidence limits show that the true value for the case lies within these limits with 95% confidence and we find that in all the 10 cases the actual LVMPI is well within the confidence interval.

The average of these residuals is usually close to zero which means that positive and negative errors cancel each other out and lead to a state of *no error* on an average. But the variance of these residuals is important. The larger the variance, the poorer the *predictive ability* of the model. In the present case the mean of the residuals is 0 and the variance is 0.0734. This can be found from the *residual statistics* which is a part of the output.

Thus, before using the regression model for prediction, one has to deeply study the nature of residuals. It is not enough, if the mean of the residuals is zero but the distribution of residuals is equally important.

Some methods for normality checks are discussed in the following section.

---

## 6.5   Quality of Residuals

Another performance indicator of a good regression model is the statistical pattern of the residuals. When the model is good, we expect the residuals follow normal distribution with mean zero and a low standard deviation.

This can be verified by plotting the standardized residuals as a normal P-P plot as shown in Figure 6.3. If all the points fall on the straight line, then the residuals can be considered as following normal distribution.

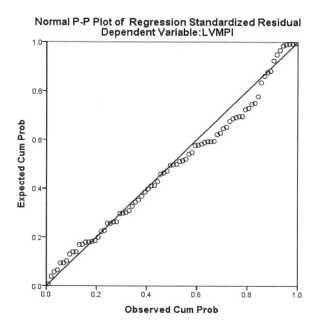

FIGURE 6.3: P-P plot of residuals.

Alternatively we can use a non-parametric test like the one-sample Kolmogorov–Smirnov test available in SPSS to test the hypothesis of normality. The null hypothesis assumes that the residuals follow normal distribution. If the p-value of the test is more than $\alpha$ (say 0.05) we accept the null hypothesis that the data is normal.

In the present case of LVMPI data, the one-sample Kolmogorov–Smirnov test shows the following results.

1. Null Hypothesis: Test distribution is normal.

2. Mean = 0.000.

3. Std. Deviation = 0.9743.

4. Asymptotic Significance (p-value) = 0.092.

Since the p-value is larger than 0.05 we accept that the distribution of standardized residuals is normal.

In the following section we discuss a method to fit regression for a selected subset of records.

---

## 6.6   Regression Model with Selected Records

It is also possible to run the complete regression model with a subset of data records and compare the resulting models.

For instance in the LVMPI data we have used MetS as an explanatory variable for LVMPI. Suppose we want to study the influence of all other variables except MetS and see separately how the regression works when the MetS is absent or present.

We can select a set of specific records by using the option *selection variable* and push MS into that box. This gives the following window to specify the rule as shown in Figure 6.4.

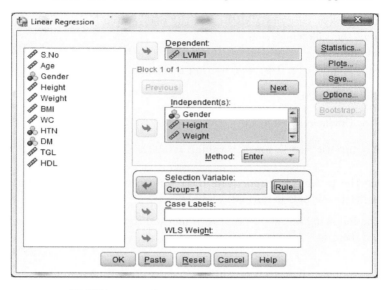

FIGURE 6.4: Options for subset regression.

The option 'Rule' shows a window in which we can specify the selection . For instance 'Group equal to 1' is the section to pick up the subset of all records for MetS = 1.

a) When MetS is taken as '1' it means "Yes" and the regression model gives $R^2$ =0.429, F = 5.764 and p = 0.020 and the variables entered into the model are Gender and WC.

b) When MetS is taken as '0' to mean 'No' and the regression model gives $R^2$ =0.366, F = 4.329 and p = 0.047 and the only variable entered into the model is Age.

It means gender and WC would explain about 43% of LVMPI behavior among those having MetS. In the other case with no MetS, it was only 'Age' that could explain the behavior of LVMPI but it is not a strong relationship.

We can also select records satisfying conditions like BMI > 22.5 and observe how the relationship is explained by the regression model.

We end this chapter with the observation that fitting a multiple linear regression model is a data-driven exercise. The output has to be interpreted with statistical arguments in a way that is better than a layman's view.

# Summary

Regression is a statistical procedure used to estimate the functional relationship between predictor variables and a response variable. This is a *cause and effect* method. The correlation coefficient between one variable and the joint effect of several other variables is called *Multiple Correlation Coefficient*, denoted by R. The adequacy of the fitted regression model is measured in terms of $R^2$ and its significance is tested by F-test. The statistical significance of regression coefficient is tested by t-test. SPSS reports additional constants called *beta coefficients* corresponding to each predictor variable, which are standardized regression coefficients. Selection of variables into regression can be done by a stepwise method, which ensures that only the most relevant variables will be included in the model. The regression model can be used to predict the Y value for given values of X variable(s). The errors in prediction are called residuals and can be analyzed with the help of residual analysis available in both MS-Excel and SPSS. PP-plot can be used to test for the normality of the observed residuals.

# Do it yourself (Exercises)

**6.1** The following data is related to 4 anthropometric measurements on the molar teeth of individuals observed in a study. It is desired to predict the age of the individual based on these measurements.

| Age (Y) | Right Molar 1 | Right Molar 2 | Left Molar 1 | Left Molar 2 |
|---------|---------------|---------------|--------------|--------------|
| 37 | 8.00 | 5.00 | 9.00 | 4.25 |
| 49 | 4.75 | 6.25 | 8.00 | 6.25 |
| 19 | 1.00 | 1.50 | 3.75 | 1.50 |
| 61 | 9.00 | 9.00 | 9.00 | 9.00 |
| 45 | 5.50 | 8.00 | 8.00 | 3.00 |
| 56 | 5.00 | 3.00 | 9.00 | 9.00 |
| 27 | 4.50 | 3.75 | 5.50 | 8.00 |
| 16 | 1.50 | 0.50 | 0.50 | 0.50 |
| 35 | 5.00 | 4.25 | 6.00 | 3.00 |
| 46 | 4.25 | 4.00 | 4.25 | 2.00 |
| 57 | 9.00 | 9.00 | 9.00 | 9.00 |
| 50 | 8.00 | 5.00 | 9.00 | 5.50 |
| 60 | 9.00 | 9.00 | 5.00 | 5.25 |
| 60 | 5.50 | 6.00 | 9.00 | 8.00 |
| 16 | 0.75 | 0.25 | 0.25 | 0.25 |
| 35 | 5.00 | 3.50 | 6.00 | 3.00 |
| 21 | 4.00 | 3.00 | 4.00 | 3.00 |
| 26 | 8.00 | 3.00 | 8.00 | 3.25 |
| 43 | 8.00 | 5.00 | 5.00 | 3.00 |
| 23 | 1.00 | 1.50 | 1.00 | 1.00 |

Construct a multiple linear regression model with a stepwise option and draw conclusions. Comment on the relative importance of predictor variables with the help of beta coefficients.

**6.2** From the following data derive a multiple linear regression model of Y (Achievement score in Mathematics) in terms of the three psychological scores (Emotional, Motivational and Parental) given.

| Emotional | Motivation | Parental | Y |
|-----------|------------|----------|-----|
| 136 | 43 | 106 | 78 |
| 147 | 46 | 113 | 68 |
| 122 | 46 | 103 | 42 |
| 144 | 48 | 107 | 77 |
| 140 | 52 | 108 | 90 |
| 153 | 52 | 121 | 80 |
| 129 | 46 | 106 | 28 |
| 162 | 54 | 118 | 74 |
| 165 | 54 | 120 | 95 |
| 145 | 56 | 122 | 87 |
| 149 | 50 | 112 | 88 |
| 152 | 46 | 134 | 89 |
| 161 | 52 | 128 | 82 |
| 155 | 46 | 85 | 61 |
| 165 | 48 | 96 | 80 |
| 129 | 48 | 145 | 74 |
| 118 | 32 | 138 | 67 |
| 159 | 52 | 108 | 91 |
| 140 | 50 | 114 | 83 |

**6.3** Practitioners often calculate correlation coefficient of Y with all X-variables. Then those X variables having significant correlation with Y are taken to fit a multiple regression model. Try this method for the data used in Illustration 6.2. Compare the resulting model with the one obtained by using the stepwise method.

**6.4** Reconsider the data given in Illustration 6.1. Using MS-Excel, calculate the predicted BMI for different values of WC, using the regression model. Also check whether the errors follow normal distribution.

**6.5** In Exercise 6.3, check in MS-Excel for the $R^2$-value when the option *intercept is zero* is selected. Is this option meaningful in this case? What is the meaning of the intercept in the model BMI = -5.9304+0.3696*WC? (Hint: if we set WC = 0, we get BMI = -5.9304. Both are incorrect. WC = 0 means no WC at all. If BMI is to be positive there will be a minimum WC. Try to get this value.)

**6.6** The choice of setting *intercept zero* is important in the regression model. Use the SPSS options and find out the difference in the results with and without this option.

## Suggested Reading

1. Johnson, R. A., & Wichern, D. W. 2014. *Applied multivariate statistical analysis*, 6$^{th}$ ed. Pearson New International Edition.

2. Bhuyan K.C. 2005. *Multivariate Analysis and Its Applications*. Kolkata: New Central Book Agency (P) Ltd.

3. Eric Vittinghoff, Stephen C. Shiboski, David V. Glidden and Charles E. McCulloch. 2004, Regression Methods in Biostatistics, Springer

4. Alvin C.Rencher, William F. Christensen 2012. *Methods of Multivariate Analysis*. 3$^{rd}$ ed. Brigham Young University: John Wiley & Sons.

# Chapter 7

# Classification Problems in Medical Diagnosis

| | | |
|---|---|---|
| 7.1 | The Nature of Classification Problems | 135 |
| 7.2 | Binary Classifiers and Evaluation of Outcomes | 137 |
| 7.3 | Performance Measures of Classifiers | 138 |
| 7.4 | ROC Curve Analysis | 144 |
| 7.5 | Composite Classifiers | 148 |
| 7.6 | Biomarker Panels and Longitudinal Markers | 149 |
| | Summary | 150 |
| | Do it yourself (Exercises) | 150 |
| | Suggested Reading | 151 |

Statistics is the science of learning from experience.

Bradley Efron (1938 – )

## 7.1 The Nature of Classification Problems

Quite frequently in medical diagnosis, the researcher is interested in *discriminating* between patients with the presence or absence of a health condition or state. e.g., the flu.. . These two states are dichotomous, often called *test positive* or *test negative* respectively and the distinction is based on one or more predictive factors, called *biomarkers*. The outcome (health condition) may also have more than two states like the stage of a cancer or severity of a disease. Using a biomarker, we wish to classify an individual into a positive or negative group. The focal point is discriminating between positives and negatives using one or more biomarkers.

In a more general context, the problem of **discrimination** is addressed by proposing a formula or procedure that can distinguish between the subjects of one group and the other. This is called *discriminant analysis* and has two objectives as given below.

a) To derive a rule of discrimination between objects of different populations (groups) based on the sample data obtained on the predictors (markers). This is also called the problem of **separation**.

b) To allocate a new subject (whose group is unknown) to one of the predefined groups with minimum error. This is called a **classification problem**.

Discrimination is therefore a process of separation of subjects, from a mixed collection and allocating them to *known groups*. This method is also known as *supervised learning* in the context of *machine learning* methods. The statistical tool to perform this is aimed at providing a classification formula that minimizes the number of *misclassifications*. In medical diagnosis we may denote the groups as Healthy (H) or Diseased (D) basing on the result of a test. Since the classification is made into two categories, it is called *binary classification*.

Sometimes there will not be predefined groups at all. We then create groups called *clusters* in such a way that all the subjects within a cluster are *similar* in some sense, whereas subjects between clusters will be *dissimilar*. This method is called *unsupervised learning*. Cluster Analysis is a multivariate tool used to address this problem. This method has several applications in social and preventive medicine, epidemiology as well as in marketing of services.

Biomarkers are also called *classifiers*. The performance of a classifier is judged in terms of its ability to correctly classify the subjects into a group. Among several possible classifiers we select the one that has the highest ability to distinguish between positive and negative categories.

A single classifier may poorly classify the subjects and sometimes there may exist a *combination* of several markers, in the form of a *composite classifier* which may perform better than the individual classifiers in predicting the correct category or group. Linear Discriminant Analysis (LDA) and Logistic Regression are two commonly used methods to handle classification problems with multiple markers.

In the following section, we discuss some standard measures of performance of classifiers and their application. We use the words *test* and *biomarker* interchangeably to indicate a *classifier*.

## 7.2 Binary Classifiers and Evaluation of Outcomes

Some classifiers are measured on a continuous scale like blood pressure, bone mineral density or serum cholesterol, and the classification is based on the comparison of the observed value with a critical value called *threshold*.

It is possible that a test shows *positive* whereas the true diagnosis may show a *negative* status. If the test result exceeds a *cutoff* value, the patient is classified into D group and treatment is started, otherwise kept in H group. If the test is not a good indicator of the true condition, this type of *misclassification* occurs.

In some cases the classifier is measured on an ordinal scale like the opinion of a radiologist about a scan with the outcome expressed as a number indicating normal, moderate, severe and very severe states of a condition. There can also be a nominal scale on which the test result is expressed, as in the case of hypertension recorded as *yes* or *no*.

In practice the test result is based on sample data and hence the result is considered as a *random variable* and theory of probability helps in making judgment on the efficiency of the classifier.

In reality, classification rules often lead to misclassification unless the rule is perfect. For instance the Oral Glucose Tolerance Test (OGTT) is considered a perfect test to detect Diabetes Mellitus (DM). Such a test is known as a *gold standard*. Similarly the IFN-$\gamma$ test is considered the gold standard to diagnose TB. Every test will be specified with a cutoff such that when the observed value exceeds the cutoff it is identified as positive; else negative.

Consider a biomarker for which there is a cutoff, which is widely accepted, tested and offers error-free classification. However, due to various reasons, this test may be either costly or not available always. For this reason clinical researchers look for alternative markers or tests that are practically feasible, cost-effective and perform close to the gold standard.

Let the test result (discrete or continuous random variable) be denoted by X and x be the observed value of X. Let *test positive* indicate a state of having disease (D) and test negative indicate healthy state (H). Let us define a cutoff value (c) such that a patient is classified into D group if $x > c$ and into H group otherwise. Suppose this method is adopted on 'n' patients and the test result is compared with the true diagnosis. The findings can be categorized as follows.

1. True Positive (TP): Count of cases where both diagnosis and test are positive,

2. False Positive (FP): Count of cases where the diagnosis is negative but test is positive,

3. True Negative (TN): Count of cases where both diagnosis and test are negative,

4. False Negative (FN): Count of cases where the diagnosis is positive but test is negative,

These counts are often shown as a matrix (two-way table) shown below. We may note that TP+ FN+FP+ TN = n.

<table>
<tr><td></td><td colspan="3">**Test result**</td></tr>
<tr><td>Diagnosis</td><td>Positive</td><td>Negative</td><td>Total</td></tr>
<tr><td>Positive</td><td>TP</td><td>FN</td><td>TP+ FN</td></tr>
<tr><td>Negative</td><td>FP</td><td>TN</td><td>FP+ TN</td></tr>
<tr><td>Total</td><td>TP+ FP</td><td>FN+ TN</td><td>n</td></tr>
</table>

In an ideal situation one expects FP $=$ FN $= 0$ but this never happens, unless the test is perfect in some sense. Some important indicators of the test performance are given in the following section.

---

## 7.3    Performance Measures of Classifiers

The performance of a classifier is assessed in terms of summary statistics (counts and proportions) with respect to a given cutoff. We also use a graphical method to display the performance along with a numerical summary of information contained in the graph. These are discussed below in detail assuming that the marker has a cutoff $c$.

**Sensitivity ($S_n$):**

It is the conditional probability of having a positive test among the patients who have a positive diagnosis (condition) and denoted by $S_n = P[X > c \mid D]$. This probability is estimated from sample data as $\widehat{S}_n = \dfrac{TP}{TP + FN}$.

This is also known as *True Positive Rate* (TPR) or *True Positive Fraction* (TPF). For a given test, if the sensitivity is 0.90 it means that in 90% of the cases where the disease is present, the test shows positive. Hence we need a high sensitivity for a test. In the context of data mining, sensitivity is referred to as *recall*. Screening tests often require high sensitivity.

### Specificity ($S_p$):

It is the conditional probability of having a negative test among the patients who have a negative diagnosis (condition) and denoted by $S_p = P[X \leqslant c \mid D]$. This probability is estimated from sample data as $\widehat{S}_p = \dfrac{TN}{TN + FP}$.

This is also known as True Negative Rate (TNR) or True Negative Fraction (TNF). A specificity of 0.80 means that in 80% of the cases where the disease is absent, the test also shows negative. In confirmatory tests, we often need a high specificity.

A good diagnostic test is supposed to have high sensitivity with reasonably high specificity. Both $S_n$ and $S_p$ values lie between 0 and 1.

### Disease prevalence:

The % of individuals who were tested positive out of those at risk, is called the *prevalence rate*. In a sample data this value is equal to $P = \dfrac{TN + FP}{N}$. The prevalence adjusted values of $S_n$ and $S_p$ are given by

$$S'_n = \frac{pS_n}{pS_n + (1 - p)(1 - S_p)} \text{ and } S'_p = \frac{(1 - p)S_p}{(1 - p)S_p + p(1 - S_n)}$$

These are useful while working with specific populations having a disease.

### Positive Predictive Value (PPV):

It is the probability that the disease is present when the test result shows positive. This is computed as $PPV = \dfrac{TP}{(TP + FP)}$.

Suppose this value is 0.85. It means that when the test result is positive there is 85% chance that the diagnosis also shows positive. This is also known as *precision*. It is also called *post-test probability* (of positive).

### Negative Predictive Value (NPV):

It is the probability that the disease is absent when the test result shows negative. This is computed as $NPV = \dfrac{TN}{(TN + FN)}$.

Suppose this value is 0.95. It means that when the test result is negative, there is 90% chance that the diagnosis also shows negative.

### Positive Likelihood Ratio ($LR^+$):

This is the ratio of the probability of a positive test result, given that the disease is present to the probability of a positive test result given that the disease is absent. We denote this by $LR^+ = \dfrac{S_n}{(1 - S_p)}$.

If $LR^+ = 5.7$, it means that an individual is 5.7 times more likely to test positive (when the disease is really present) when compared to those who do not have the disease. Values of $LR^+$ which are less than 1 are usually not considered for any comparison.

### Negative Likelihood Ratio ($LR^-$):

This is the ratio of the probability of a negative test result given that the disease is present to the probability of a negative test result given that the disease is absent.

We compute this as $LR^- = \dfrac{(1 - S_n)}{S_p}$.

Again this value indicates the likelihood of obtaining a false negative compared to those for whom the disease is really absent.

Suppose there is a marker for which the actual cutoff value is not known but different options are available like $c_1, c_2, \ldots, c_k$. At each cutoff we get a pair of values $(S_n, S_p)$ from which, PPV, NPV, $LR^+$ and $LR^-$ can be calculated.

In the following section let us understand how these values are calculated from a real data. Consider the following illustration.

**Illustration 7.1** Acute Physiology and Chronic Health Evaluation (APACHE) score and Sepsis Related Organ Failure Assessment (SOFA) score are two commonly used scoring systems to assess the status of a patient after admission into the Intensive Care Unit (ICU) of a hospital. These scores are calculated based on the patient's parameters measured within 24/48 hours after admission into the ICU and a higher score predicts death. The variables and codes are described below.

| Variable | Parameter | Description |
|---|---|---|
| X1 | Outcome | End event for the patient (1 = Dead, 0 = Alive) |
| X2 | Age | Age in years |
| X3 | Gender | 1 = Male, 0 = Female |
| X4 | Shock | 1 = Yes, 0 = No Shock |
| X5 | Diagnosis | 1 = Sepsis, 2 = Severe Sepsis and 3 = Septic Shock |
| X6 | Ventilator | Ventilator required or not (1 = Required, 0 = Not required) |
| X7 | AKI | Acute Kidney Infection (1 = Yes, 0 = No) |
| X8 | APACHE | Acute Physiological and Chronic Health Evaluation score |
| X9 | SOFA | Sepsis Related Organ Failure Assessment score |
| X10 | DurHospStay | Duration of Hospital Stay (days) |
| X11 | Days of Vent | Days sent on ventilator |
| X12 | SCr | Serum Creatine |
| X13 | SUrea | Serum Urea |

Table 7.1 contains a portion of the data with 20 records. For further reference this data will be called *ICU scores data*.

TABLE 7.1: ICU scores data with all variables

| S.No | X1 | X2 | X3 | X4 | X5 | X6 | X7 | X8 | X9 | X10 | X11 | X12 | X13 |
|------|----|----|----|----|----|----|----|----|----|-----|-----|------|------|
| 1 | 0 | 20 | 0 | 3 | 1 | 1 | 0 | 12 | 4 | 12 | 5 | 0.63 | 24 |
| 2 | 1 | 52 | 1 | 2 | 0 | 1 | 1 | 21 | 16 | 12 | 5 | 3.51 | 197 |
| 3 | 0 | 25 | 1 | 2 | 0 | 0 | 1 | 20 | 12 | 9 | 0 | 3.60 | 225 |
| 4 | 0 | 53 | 1 | 2 | 1 | 1 | 0 | 20 | 8 | 10 | 3 | 1.80 | 93 |
| 5 | 0 | 15 | 1 | 2 | 1 | 1 | 0 | 15 | 10 | 14 | 5 | 1.95 | 117 |
| 6 | 0 | 40 | 1 | 3 | 1 | 1 | 1 | 21 | 12 | 10 | 4 | 3.50 | 59 |
| 7 | 0 | 70 | 1 | 2 | 0 | 1 | 1 | 16 | 12 | 7 | 2 | 3.38 | 148 |
| 8 | 1 | 50 | 1 | 2 | 1 | 1 | 1 | 22 | 13 | 12 | 5 | 4.65 | 157 |
| 9 | 0 | 27 | 1 | 3 | 1 | 0 | 0 | 9 | 9 | 9 | 0 | 1.42 | 46 |
| 10 | 0 | 30 | 0 | 2 | 1 | 1 | 0 | 13 | 5 | 22 | 18 | 0.60 | 39 |
| 11 | 1 | 47 | 1 | 3 | 1 | 1 | 1 | 23 | 17 | 12 | 12 | 5.29 | 109 |
| 12 | 0 | 23 | 0 | 2 | 0 | 1 | 0 | 15 | 7 | 8 | 4 | 0.93 | 41 |
| 13 | 0 | 19 | 0 | 2 | 0 | 0 | 0 | 8 | 5 | 10 | 0 | 0.28 | 26 |
| 14 | 0 | 40 | 0 | 2 | 0 | 1 | 0 | 11 | 8 | 9 | 4 | 0.89 | 38 |
| 15 | 1 | 30 | 0 | 3 | 1 | 1 | 1 | 21 | 17 | 4 | 4 | 2.29 | 89 |
| 16 | 0 | 27 | 0 | 2 | 0 | 0 | 0 | 14 | 8 | 8 | 0 | 0.33 | 39 |
| 17 | 1 | 39 | 0 | 2 | 1 | 1 | 1 | 32 | 18 | 12 | 10 | 4.08 | 82 |
| 18 | 0 | 27 | 1 | 2 | 1 | 1 | 0 | 10 | 7 | 11 | 7 | 0.49 | 35 |
| 19 | 0 | 16 | 0 | 2 | 1 | 1 | 1 | 13 | 10 | 33 | 24 | 2.04 | 56 |
| 20 | 0 | 63 | 0 | 2 | 0 | 1 | 0 | 23 | 8 | 15 | 4 | 0.16 | 115 |

*(Data courtesy: Dr Alladi Mohan, Department of Medicine, Sri Venkateswara Institute of Medical Sciences (SVIMS), Tirupati.)*

Let us consider the variable Outcome, SOFA, SCr and SUrea. We wish to determine how sensitive SOFA is to detect outcome and what could be the best cutoff to distinguish between alive and dead. The data analysis is however done on 50 records out of 248 from the original data.

**Analysis:**

Since no cutoff is given, let us start with the simple average of the SOFA values which come to 9.26 as a possible cutoff. Then, for a patient *death* will be predicted if the SOFA > 9.26; else *death* not predicted. The calculation of $S_n$ and $S_p$ basically requires the count of values from the actual data which can be done with an MS-Excel sheet.

The True Positive (TP) count of records in the data for which both the conditions i) SOFA > 0.9.26 and ii) Outcome = 1 are true, can be found with the MS-Excel function COUNTIFS( ). By changing the conditions we can find the remaining FN, FP and TN values. After finding the marginal totals, we can determine $S_n$ and $S_p$. Figure 7.1 is a template in MS-Excel to derive these numbers. The template is interactive, in the sense that when the cutoff value is changed, the corresponding results are automatically changed.

$f_x$     =COUNTIFS($C$2:$C$51,$J$4,$B$2:$B$51,"=1")

| G | H | I | J | K |
|---|---|---|---|---|
| | | Cutoff | Criterion | |
| | | 12.00 | >12 | |
| | | Positive | Negative | |
| | Case | 7 | 5 | 12 |
| | Control | 0 | 38 | 38 |
| | | 7 | 43 | 50 |

FIGURE 7.1: MS-Excel template to calculate the sensitivity and specificity.

We find that with a cutoff SOFA $> 9.26$, the sensitivity is 91.7% while the specificity is 63.2%. Suppose the cutoff is changed. We only need to change the value in the cell N4 and press the 'enter' key. With two different cutoff values, we get the results as shown in Table 7.2.

TABLE 7.2: Changes in sensitivity and specificity due to changes in the cutoff

| **SOFA > 10** | | | | **SOFA > 12** | | | |
|---|---|---|---|---|---|---|---|
| Diagnosis | Test +ve | Test -ve | Total | Diagnosis | Test +ve | Test -ve | Total |
| Case | 10 | 2 | 12 | Case | 7 | 5 | 12 |
| Control | 6 | 32 | 38 | Control | 0 | 38 | 38 |
| Total | 16 | 34 | 50 | Total | 7 | 43 | 50 |
| $S_n = 0.833, S_p = 0.842$ | | | | $S_n = 0.583, S_p = 1.000$ | | | |

When the cutoff is increased to 10 the sensitivity has decreased to 83.3% but the specificity increased to 84.2%. It is therefore necessary to balance these indices and find the best cutoff.

Again when the cutoff is increased from 10 to 12 the sensitivity has decreased to 58.3% while the specificity has increased to 100% (not shown as a table).

The Receiver Operations Characteristic (ROC) curve is a graphical description of the sensitivity and specificity at different possible cutoff values of a marker. It is a plot of $S_n$ against $(1-S_p)$. We can calculate $S_n$ and $S_p$ at different values of SOFA as possible cutoffs, as shown in Table 7.3.

TABLE 7.3: $S_n$ and $(1-S_p)$ values at different possible cutoff values on SOFA

| SOFA | $S_n$ | $S_p$ | $1-S_p$ |
|------|-------|-------|---------|
| >4   | 1.000 | 0.105 | 0.895   |
| >5   | 1.000 | 0.211 | 0.789   |
| >6   | 1.000 | 0.263 | 0.737   |
| >7   | 1.000 | 0.368 | 0.632   |
| >8   | 0.917 | 0.605 | 0.395   |
| >9   | 0.917 | 0.632 | 0.368   |
| >10  | 0.833 | 0.842 | 0.158   |
| >11  | 0.750 | 0.868 | 0.132   |
| >12  | 0.583 | 1.000 | 0.000   |
| >13  | 0.333 | 1.000 | 0.000   |
| >16  | 0.250 | 1.000 | 0.000   |
| >17  | 0.083 | 1.000 | 0.000   |
| >18  | 0.000 | 1.000 | 0.000   |

The ROC curve is obtained by a scatter plot of $S_n$ versus $(1-S_p)$ which can be plotted using MS-Excel as shown in Figure 7.2. We observe that the entire graph lies within a unit square because both $S_n$ and $(1-S_p)$ can have a maximum area of 1 (unity).

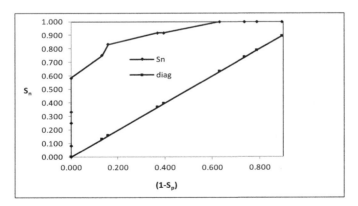

FIGURE 7.2: ROC curve with diagonal line drawn using MS-Excel.

In an ideal situation, one would like to have a cutoff which produces the highest sensitivity as well as specificity. If specificity is low there would be more *false negatives* and with low sensitivity we get more *false positives*. The solution to balance these two measures is a single intrinsic measure called the *Area Under the Curve (AUC)*.

Suppose at each cutoff, $S_n$ is equal to $(1-S_p)$. Then the ROC curve becomes a straight line, called a *diagonal line*, shown in Figure 7.1 and the area below it is 0.5. If the sample ROC curve exactly matches with the diagonal line, it means that the corresponding marker has only 50% chance of distinguishing between +ve and –ve instances (alive and death) and has no relevance. We therefore need to find markers for which the AUC is above 0.5 and also as high as possible.

The AUC can be calculated with MS-Excel using the Trapezoidal rule but there are software tools to work out this number easily. The AUC can be nicely interpreted as a probability, regarding the ability of the marker to distinguish between the two groups.

In general define X as the test score of a randomly selected *diseased* person and Y as the test score of a randomly selected *healthy* person. Then the AUC denotes the probability that a randomly selected individual having the disease will have X value higher than the Y value of the randomly selected individual without disease. This is expressed a AUC= $P(X > Y)$.

For instance, if the AUC of a marker is 0.94 it means that in 94% of situations, the marker can correctly determine the group to which the person belongs to.

Hence we need not consider markers having AUC smaller than 0.5 because it is no better than a *classification by chance*.

In the following section a detailed methodology of the ROC curve is discussed.

---

## 7.4   ROC Curve Analysis

By constructing and analyzing the ROC curve from sample data, we can arrive at useful conclusions about the efficiency of biomarkers and draw inferences about their true characteristics.

We can use a software package like SPSS or MedCalc to construct a ROC curve and to derive all metrics like sensitivity, specificity, AUC. When the marker is measured on a continuous scale, every value can be considered as a possible cutoff and we can pick the optimal one. Kronozowki and Hand (2009) provided a detailed theory for the ROC curve analysis with continuous data.

Consider the following illustration.

**Illustration 7.2** Reconsider the data discussed used Illustration 7.1. Let us use *MedCalc v15.2* and choose ROC curve analysis.

The data from either MS-Excel or SPSS can be opened in this software. The input and output options are shown in Figure 7.3.

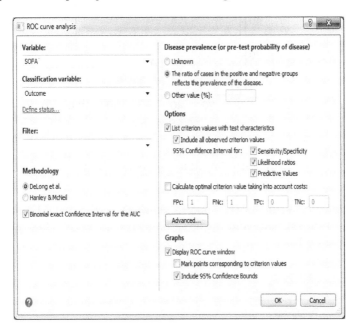

FIGURE 7.3: Input and output options for creating a ROC curve using Med-Calc.

The ROC curve is shown in Figure 7.4, as a dark line and the 95% confidence interval for the curve is also displayed as two lines above and below the curve. The curve is well above the diagonal line and hence the marker SOFA seems to have good ability to distinguish between alive and dead.

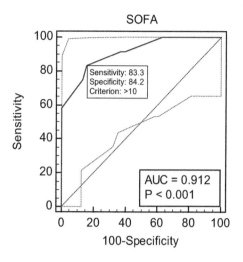

FIGURE 7.4: ROC curve for SOFA.

The summary results of the analysis shows the estimated AUC, the standard error of the estimate and the 95% confidence intervals. These details are shown in Figure 7.5.

| Variable | SOFA |
|---|---|
| Classification variable | Outcome |

| | |
|---|---|
| Sample size | 50 |
| Positive group [a] | 12 (24.00%) |
| Negative group [b] | 38 (76.00%) |

[a] Outcome = 1
[b] Outcome = 0

| | |
|---|---|
| Disease prevalence (%) | 24.0 |

**Area under the ROC curve (AUC)**

| | |
|---|---|
| Area under the ROC curve (AUC) | 0.912 |
| Standard Error [a] | 0.0487 |
| 95% Confidence interval [b] | 0.798 to 0.974 |
| z statistic | 8.474 |
| Significance level P (Area=0.5) | <0.0001 |

[a] DeLong et al., 1988
[b] Binomial exact

FIGURE 7.5: Statistical summary of ROC curve analysis.

It is can be seen that AUC = 0.912 with 95% CI [0.798, 0.974]. The analysis provides a statistical test for the significance of the observed AUC by using a Z-test. The null hypothesis is that the true AUC is 0.5 (50% chance of misclassification). The p-value is <0.0001 and hence a sample AUC of 0.912 is not an occurrence by chance and therefore significant.

TABLE 7.4: Possible cutoff values of SOFA and the optimal cutoff

| Criterion | Sensitivity | 95% CI | Specificity | 95% CI | +LR | -LR |
|---|---|---|---|---|---|---|
| ⩾2 | 100.00 | 73.5 - 100.0 | 0.00 | 0.0 - 9.3 | 1.00 | |
| >2 | 100.00 | 73.5 - 100.0 | 5.26 | 0.6 - 17.7 | 1.06 | 0.00 |
| >3 | 100.00 | 73.5 - 100.0 | 7.89 | 1.7 - 21.4 | 1.09 | 0.00 |
| >4 | 100.00 | 73.5 - 100.0 | 10.53 | 2.9 - 24.8 | 1.12 | 0.00 |
| >5 | 100.00 | 73.5 - 100.0 | 21.05 | 9.6 - 37.3 | 1.27 | 0.00 |
| >6 | 100.00 | 73.5 - 100.0 | 26.32 | 13.4 - 43.1 | 1.36 | 0.00 |
| >7 | 100.00 | 73.5 - 100.0 | 36.84 | 21.8 - 54.0 | 1.58 | 0.00 |
| >8 | 91.67 | 61.5 - 99.8 | 60.53 | 43.4 - 76.0 | 2.32 | 0.14 |
| >9 | 91.67 | 61.5 - 99.8 | 63.16 | 46.0 - 78.2 | 2.49 | 0.13 |
| >10 | 83.33 | 51.6 - 97.9 | 84.21 | 68.7 - 94.0 | 5.28 | 0.20 |
| >11 | 75.00 | 42.8 - 94.5 | 86.84 | 71.9 - 95.6 | 5.7 | 0.29 |
| >12 | 58.33 | 27.7 - 84.8 | 100.00 | 90.7 - 100.0 | | 0.42 |
| >13 | 33.33 | 9.9 - 65.1 | 100.00 | 90.7 - 100.0 | | 0.67 |
| >16 | 25.00 | 5.5 - 57.2 | 100.00 | 90.7 - 100.0 | | 0.75 |
| >17 | 8.33 | 0.2 - 38.5 | 100.00 | 90.7 - 100.0 | | 0.92 |
| >18 | 0.00 | 0.0 - 26.5 | 100.00 | 90.7 - 100.0 | | 1.00 |

The next important aspect of ROC analysis is the determination of the optimal cutoff. Table 7.4 shows different possible values as criterion and finds the sensitivity and specificity at each value. For instance with a criterion of '>6' we have $S_n = 100\%$ and $S_p = 26.32\%$ (both $S_n$ and $S_p$ are expressed as percentages for the sake of interpretation).

There are different methods of finding the optimal cutoff. One method is to calculate the value of $(S_n + S_p - 1)$ at each possible value and pick up the largest; the corresponding criterion is the optimal cutoff. Another method is based on finding the criterion value for which the line joining the top right corner with the ROC has the smallest distance from the curve.

Alternatively, one can use the value having the longest vertical distance from the curve to the horizontal axis, which is also known as the *Youden index*. Software packages use different algorithms to find the best optimal point.

The optimal cutoff is '> 10' associated with the Youden index (J= 0.6754) with AUC = 0.912 with 83.33% sensitivity and 84.21% specificity. The confidence intervals for sensitivity and specificity are also shown in the Table 7.4 along with +ve and -ve LR values.

In the following section we compare the performance of several biomarkers in distinguishing between the two outcomes (alive/dead). We use the same data and carry out ROC analysis using MedCalc.

**Illustration 7.3** Reconsider the data discussed in Illustration 7.1. Suppose we consider all three markers SOFA, SCr and SUrea. We wish to know how good they are in predicting the outcome. Applying the ROC analysis to these variables we get the results shown in Table 7.5.

TABLE 7.5: Comparison of several biomarkers

| Marker | Cutoff | Sensitivity | Specificity | +LR | -LR | AUC | p-value |
|--------|--------|-------------|-------------|-----|-----|-----|---------|
| SOFA   | >10    | 83.33%      | 84.21%      | 5.28 | 0.20 | 0.912 | <0.0001 |
| SCr    | >3.5   | 50.00%      | 92.11%      | 6.33 | 0.54 | 0.739 | 0.0055  |
| SUrea  | >77    | 58.33%      | 76.32%      | 2.46 | 0.55 | 0.651 | 0.0982  |

In terms of AUC, none of the above markers is more promising than the SOFA and the AUC produced by them is not significant (to mean that this much of AUC could even appear due to chance!) for SUrea. Hence SOFA is the best classifier among the three.

In some cases a single marker may not be able to properly distinguish between cases and controls. We may consider combining two or more markers to create a new marker.

In the following section we examine composite classifications and their use.

## 7.5   Composite Classifiers

Sometimes there could be several independent markers for the same disease, each having its own AUC. For instance, in addition to CIMT, variables like age, BMI, TGL etc., may show a combined influence on the classification and probably reduce the percentage of misclassifications. It is therefore worth considering a multivariate approach to propose new classifiers whose performance is better than any of the univariate classifiers.

Let $X_1, X_2,\ldots,X_p$ be a set of p-markers each of which is a possible candidate to classify the subjects. One way of combining them is to define a new function $Z = f(X_1, X_2,\ldots,X_p)$ and use the observed Z as a new marker called *composite marker* or *composite classifier*.

We can also perform ROC analysis and propose a new classification rule.

One simple method is to define a linear model of the form

$$Z = \beta_0 + \beta_1 X_1 + \beta_2 X_2 + \ldots + \beta_p X_p$$

where $\beta_0, \beta_1, \ldots, \beta_p$ are weights (regression coefficients) given to the corresponding test values.

These weights can be determined by using a method called *multiple logistic regression*. Such data is often called the *training data*. When a new individual arrives, we can compute Z and classify the individual as a case if $Z > Z^*$ or control otherwise.

Several researchers recently have studied the utility of biomarkers panels instead of a single individual marker. A brief note on this is given in the following section.

---

## 7.6   Biomarker Panels and Longitudinal Markers

Recently, the concept of 'biomarker panel' has been widely used in the diagnosis as well as disease progression studies.

A panel is a set of biomarkers which exhibit better discriminatory power than a single maker.

Markers can be combined using a binary multiple logistic regression to yield a composite marker. Alternatively, principal component regression can be used with a dichotomous outcome variable.

Simona Mihai *et. al.* (2016) have discussed the utility of a multiplex biomarker panel in Chronic Kidney Disease (CKD). They found panels of mineral and bone disorder biomarkers to be more relevant than a single marker to detect patients in early stage CKD.

Sophie Paczesny *et.al* (2009) came up with a biomarker panel for distinguishing between GVHD +ve and -ve patients, They used logistic regression and arrived at a composite marker and derived the sensitivity and specificity basing on the score obtained from the logistic regression model.

Longitudinal markers are those used to predict disease progression. This issue arises in an area called *disease progression studies*. The dynamics of markers over a time span leads to the definition of longitudinal surrogate markers. Lawrence *et.al* (2017) provide a review of such markers in the context of Alzheimer's disease.

In the next two chapters we discuss methods of constructing composite classifiers for binary classification, using regression methods.

## Summary

Problems in medical diagnosis often deal with distinguishing between a true and false state basing on a classifier or marker. This is a binary classification and the best classifier will have no error in classification. Researchers propose surrogate markers as alternative classifiers and wish to know how well they can distinguish between cases and controls. ROC curve analysis helps to assess the performance of a classifier and also to find the optimal cutoff value. ROC analysis is popularly used in predictive models for clinical decision making. Biomarker panels and longitudinal markers are widely used as multivariate tools in ROC analysis.

## Do it yourself (Exercises)

**7.1** The following data is also a portion of data used in Illustration 7.1. All three markers are categorical.

| S.No | Outcome | Diagnosis | Shock | AKI |
|------|---------|-----------|-------|-----|
| 1 | 0 | 3 | 1 | 0 |
| 2 | 1 | 2 | 0 | 1 |
| 3 | 0 | 2 | 0 | 1 |
| 4 | 0 | 2 | 1 | 0 |
| 5 | 0 | 2 | 1 | 0 |
| 6 | 0 | 3 | 1 | 1 |
| 7 | 0 | 2 | 0 | 1 |
| 8 | 1 | 2 | 1 | 1 |
| 9 | 0 | 3 | 1 | 0 |
| 10 | 0 | 2 | 1 | 0 |
| 11 | 1 | 3 | 1 | 1 |
| 12 | 0 | 2 | 0 | 0 |
| 13 | 0 | 2 | 0 | 0 |
| 14 | 0 | 2 | 0 | 0 |
| 15 | 1 | 3 | 1 | 1 |
| 16 | 0 | 2 | 0 | 0 |
| 17 | 1 | 2 | 1 | 1 |
| 18 | 0 | 2 | 1 | 0 |
| 19 | 0 | 2 | 1 | 1 |
| 20 | 0 | 2 | 0 | 0 |

Compute the AUC for all three markers and compare the ROC curves.

**7.2** Prepare simple templates in MS-Excel to count the true positive and false positive situations by using dummy data.

**7.3** Compare the options available in MedCalc and SPSS in performing ROC curve analysis.

**7.4** The MS-Excel Add-ins 'Real-stat' has an option for ROC curve analysis. Use it on the data given in Exercise 7.1 and obtain the results.

**7.5** How do you get prevalence adjusted $S_n$ and $S_p$ from MedCalc software?

**7.6** Use the following data and find TP, FP, TN and FN on X1 using Class as outcome (0 = Married, 1 = Single). The cutoff may be taken as the mean of X1.

| S.No | 1 | 2 | 3 | 4 | 5 | 6 | 7 | 8 | 9 | 10 |
|------|---|---|---|---|---|---|---|---|---|----|
| Class | 0 | 0 | 0 | 0 | 0 | 0 | 0 | 0 | 0 | 0 |
| X1 | 20 | 19 | 41 | 16 | 45 | 75 | 38 | 40 | 19 | 21 |

| S.No | 11 | 12 | 13 | 14 | 15 | 16 | 17 | 18 | 19 | 20 |
|------|----|----|----|----|----|----|----|----|----|----|
| Class | 1 | 1 | 1 | 1 | 1 | 1 | 1 | 1 | 1 | 1 |
| X1 | 71 | 26 | 41 | 55 | 22 | 32 | 47 | 56 | 23 | 62 |

# Suggested Reading

1. Hanley. A. James and Barbara J Mc Neil. 1982. A Meaning and Use of the area under a Receiver Operating Characteristic (ROC) curves. *Radiology* 143: 29 - 36.

2. Pepe, M.S. 2000. Receiver operating characteristic methodology. *Journal of American Statistical Association* 95:308 – 311.

3. Vishnu Vardhan R and Sarma K.V.S. 2012. Determining the optimal cut-point in an ROC curve - A spreadsheet approach. *International Journal of Statistics and Analysis* 2(3): 219 - 225.

4. Krzanowski. J. Wojtek and David J. Hand. 2009. *ROC Curves for Continuous Data.*: Chapman & Hall/CRC.

5. Cohen, J.A. 1960. A coefficient of agreement for nominal scales, educational and psychological measurement. *Psychological Measurement* 20:37–46.

6. Uzay Kaymak, Arie Ben-David, and Rob Potharst. 2010. AUK: A simple alternative to the AUC. *Engineering Applications of Artificial Intelligence* 25:ERS-2010-024-LIS.

7. Simona Mihai, Elena Codrici, Ionela Daniela Popescu, Ana-Maria Enciu, Elena Rusu, Diana Zilisteanu, Radu Albulescu, Gabriela Anton, and Cristiana Tanase. 2016. Proteomic biomarkers panel: New insights in chronic kidney disease. *Dis Markers.*(http://dx.doi.org/10.1155/2016/3185232).

8. Emma Lawrence, Carolin Vegvari, Alison K. Ower, Christoforos Hadjichrysanthou, Frank de Wolf, Roy M. Anderson. 2017. A systematic review of longitudinal studies which measure Alzheimer's disease biomarkers. *Journal of Alzheimer's Disease.*59(Suppl 3): 1-21.

9. Paczesny S, Krijanovski OI, Braun TM, Choi SW, Clouthier SG, Kuick R, Misek DE, Cooke KR, Kitko CL, Weyand A, Bickley D, Jones D, Whitfield J, Reddy P, Levine JE, Hanash SM, Ferrara JL. 2009. A biomarker panel for acute graft-versus-host disease. *Blood* 113(2): 273-8.

# Chapter 8

# Binary Classification with Linear Discriminant Analysis

8.1     The Problem of Discrimination .................................. 153
8.2     The Discriminant Score and Decision Rule ..................... 155
8.3     Understanding the Output ..................................... 158
8.4     ROC Curve Analysis of Discriminant Score ................... 161
8.5     Extension of Binary Classification ........................... 163
      Summary ...................................................... 164
      Do it yourself (Exercises) ..................................... 164
      Suggested Reading ............................................ 167

Prediction is very difficult, especially about the future.

Niels Bohr (1885 – 1962)

## 8.1   The Problem of Discrimination

Discriminant Analysis (DA) is a multivariate statistical tool used to *classify* a set of objects or individuals into one of the predetermined *groups*. DA is also a tool to solve the problem of *separation* where a sample of n objects will be separated into predefined groups. We wish to develop a rule to discriminate between the objects belonging to one group and the other. The result of the classification rule then becomes a random variable and some misclassifications may occur. The objective therefore is to arrive at a rule that would minimize the percentage of misclassification.

Linear Discriminant Analysis (LDA) is a statistical procedure proposed by R.A. Fisher (1936) in which multiple linear regression is used to relate the outcome variable (Y) with several explanatory variables, each of which is a possible marker to determine Y. When the outcome is *dichotomous* (taking only two values) the classification will be binary. This is not the case with the usual multiple linear regression where Y is taken as continuous. The regression model connects Y with the explanatory variables, which can be either continuous or categorical. In LDA the dichotomous variable (Y) are regressed on to the explanatory variables as done in the context of multiple linear regression.

Let A and B be two populations (groups) from which certain characteristics have been measured. For instance A and B may represent *case* and *control* subjects. Let $X_1$, $X_2$,...,$X_p$ be the p-variables measured from samples of size $n_1$ and $n_2$ drawn from the two groups A and B respectively. All the samples belonging to A can be coded as '1' and the other coded as '2' or '0'. Obviously these two groups are mutually distinct because an object cannot belong to both A and B and there is no third group to mention.

The statistical procedure in LDA is based on 3 different methods :

1. Maximum Likelihood Discriminant Rule (ML method),

2. Bayes Discriminant Rule and

3. Fisher's Linear Discriminant Function.

While SPSS offers all the methods for computations, we confine our attention to Fisher's Linear Discriminant Function only.

Let $\mu_1$ and $\mu_2$ be the means vectors of the two multivariate populations with covariance matrices $\Sigma_1$ and $\Sigma_2$ respectively. LDA is useful only when $H_0 : \mu_1 = \mu_2$ is rejected by MANOVA. In other words, we need the mean vectors in the two groups shall differ significantly in the two populations.

The following conditions are assumed in the context of LDA (see Rencher anf Christensen (2012)).

1. The two groups have the same covariance matrix ($\Sigma_1 = \Sigma_2 = \Sigma$ ). In the case of unequal covariance matrices, we use another tool called *quadratic discriminant analysis*.

2. A normality condition is not required. When this condition is also true then the Fisher's method gives optimal classification.

3. Data should not have outliers. (If there are a few they should be handled suitably.)

4. The size of the smaller group (number of records in the group) must be larger than the number of predictor variables to be used in the model.

The LDA approach is to develop a measure called *Discriminant Score* basing on the data on X-variables and to classify an object into group A if the score exceeds a *cutoff* value and to group B otherwise. The statistical technique is to fit a *multiple linear regression* model of the type $Y = b_1X_1 + b_2X_2 + \ldots + b_pX_p$ where Y takes values 1 or 0 and b's are the weights, whose values can be estimated from the sample data. Fisher's approach is to find the discriminant score D in such a way that any *overlap* between the populations is minimized.

Since there are $n_1$ known cases in group A and $n_2$ known cases in B, it is possible to build a linear regression model using these $(n_1 + n_2)$ data points. We may use a full regression or a stepwise regression to determine the regression model.

---

## 8.2   The Discriminant Score and Decision Rule

After fitting the regression model, the regression coefficients (unstandardized) can be used to derive the discriminant score D. This new score is nothing but the predicted Y value from the regression model. For instance, a discriminant function estimated from the sample data will be like $D = -5.214 - 0.021X_1 + 0.020X_2$. Given the values of X-variables of a new individual, we can calculate the score from the regression model. Suppose $\bar{x}_{11}$ and $\bar{x}_{12}$ represents the means of $X_1$ and $X_2$ in group-1 and $\bar{x}_{21}$ and $\bar{x}_{22}$ in group-2 respectively.

Substituting these values in the discriminant function we get:

$$\overline{D}_1 = b_0 + b_1\bar{x}_{11} + b_2\bar{x}_{12} \text{ (average score from group-1 at the centroid)}$$

$$\overline{D}_2 = b_0 + b_1\bar{x}_{21} + b_2\bar{x}_{22} \text{ (average score from group-2 at the centroid)}$$

The vector $[\bar{x}_{11}, \bar{x}_{12}]$ is called the *centroid of group-1* and $[\bar{x}_{21}, \bar{x}_{22}]$ is called the *centroid of group-2*.

Define $D_0 = \frac{1}{2}(\overline{D}_1 + \overline{D}_2)$ which is the average of the score obtained at the centroids of the two groups. The classification rule is as follows:

"If $\overline{D}_1 > \overline{D}_2$, classify the subject into group-1; else classify into group-2"

Suppose it is known that an individual, selected at random, has a prior probability (chance) p of being from group-1 and (1-p) from group-2. We may use this information to classify a new individual more likely into the group having higher probability.

When individuals are classified with the above rule, the number of correct and wrong classifications can be arranged as a *classification table* like the one in Table 8.1.

The percentage of misclassifications will be $\left(\dfrac{n_{12} + n_{21}}{n_1 + n_2}\right) \times 100$ . Standard statistical software like SPSS reports the percentage of correct classifications instead of misclassifications.

TABLE 8.1: Classification by linear discriminant function

| Actual group | Predicted group | | Total |
|---|---|---|---|
| | 1 | 2 | |
| 1 | $n_{11}$ (correct) | $n_{12}$ (misclassified) | $n_1$ |
| 2 | $n_{21}$ (misclassified) | $n_{22}$ (correct) | $n_2$ |

Consider the following illustration.

**Illustration 8.1** : Reconsider the ICU scores data used in Illustration 7.1. Table 8.2 shows a portion of 20 records with two predictors SOFA and APACHE along with outcome variable (1 = Dead, 0 = Alive).

TABLE 8.2: ICU scores data with SOFA, APACHE and outcome

| S.No | Outcome | SOFA | APACHE | S.No | Outcome | SOFA | APACHE |
|---|---|---|---|---|---|---|---|
| 1 | 0 | 4 | 12 | 11 | 1 | 17 | 23 |
| 2 | 1 | 16 | 21 | 12 | 0 | 7 | 15 |
| 3 | 0 | 12 | 20 | 13 | 0 | 5 | 8 |
| 4 | 0 | 8 | 20 | 14 | 0 | 8 | 11 |
| 5 | 0 | 10 | 15 | 15 | 1 | 17 | 21 |
| 6 | 0 | 12 | 21 | 16 | 0 | 8 | 14 |
| 7 | 0 | 12 | 16 | 17 | 1 | 18 | 32 |
| 8 | 1 | 13 | 22 | 18 | 0 | 7 | 10 |
| 9 | 0 | 9 | 9 | 19 | 0 | 10 | 13 |
| 10 | 0 | 5 | 13 | 20 | 0 | 8 | 23 |

We wish to combine these two scores and propose a new classifier in the form of a composite marker using LDA.

**Analysis:**

We first read the data in SPSS and check for normality of SOFA and APACHE scores in the two groups. This is done by using the one-sample Kolmogorov - Smirnov test with the following menu options.

1. *Select cases → Outcome = 1*

2. *Analyze → Nonparametric tests → Legacy dialogs → 1-Sample K-S → Test variable list = (SOFA, APACHE) → Test distribution = Normal → OK*

This gives Kolmogorov-Smirnov statistic Z = 0.726 (p = 0.667) for SOFA and Z = 0.679 (p = 0.746) for APACHE. Since both the p-values are larger than 0.05 we find no reason for non-normality and hence normality is accepted in this group.

By selecting the cases with Outcome = 0 and repeating step-2 above we get Kolmogorov - Smirnov Z = 0.842 (p = 0.478) for SOFA and Z = 0.777 (p = 0.581) for APACHE and again the normality assumption is accepted. The two-group LDA can be performed with the help of SPSS.

The following are the steps.

1. Choose the options, Analyze → Classify → Discriminant.

2. Select Grouping variable from the list and assign the range as 0, 1 indicating the two groups (Figure 8.1).

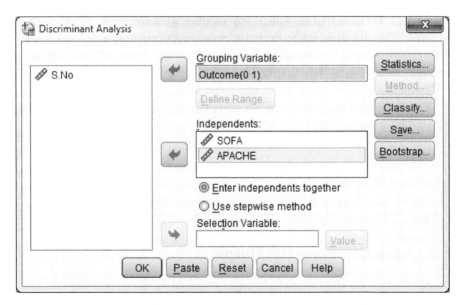

FIGURE 8.1: SPSS options for Linear Discriminant Analysis.

3. Select SOFA, APACHE as independent variables.

4.  Choose the option 'Enter independents together' for running the regression. We may also choose the stepwise method if we wish to pickup the most promising variables alone from a big list of independent variables. Since we have only two variables, we do not need the stepwise method.

5.  Click the tab 'Statistics'.

6.  Choose 'Function Coefficients as Fisher's and choose 'Unstandardised'. Press 'continue'.

7.  Click on the tab 'Classify'.

8.  Choose the prior probability as 'Compute from group size' (we may also choose the other option 'All groups equal').

9.  Choose 'Summary table'. This gives the two-way table of actual and predicted classification and reports the percentage of correct classification.

10. Click on tab 'Save'. Check now that two important items are saved into the original data file: a) Predicted membership and b) Discriminant scores.

For each data record, the predicted membership obtained by the LDA model will be saved from which the number of misclassifications can be counted. The discriminant score is the value obtained from the LDA model after substituting the values of SOFA and APACHE for each record. This is in fact the *composite score* that combines knowledge of both scores into a single index.

The distinguishing ability of each score in predicting death can be studied separately by using ROC curve analysis. Each score will have a separate cutoff value. We use this score to perform ROC analysis treating this new score as a new classifier.

## 8.3   Understanding the Output

SPSS gives all the relevant statistics required to derive the discriminant function. However, the practitioner can focus on some important values only, as discussed below.

SPSS reports canonical correlation analysis (using a method of cross-covariance matrix) to assess how many independent variables put together can explain the total variation in the data. This is expressed in terms of an index called *eigen value*. In this case the eigen value is 1.025 and it was found to explain 100% of variation between the two groups and the canonical correlation is 0.711 which was also found to be significantly different from zero.

The method therefore produces a single canonical discriminant function with coefficients given in Table 8.3.

TABLE 8.3: Discriminant function coefficients

| Variable | Coefficients | |
| --- | --- | --- |
| | Unstandardized | Standardized |
| SOFA | 0.194 | 0.551 |
| APACHE | 0.139 | 0.639 |
| (Constant) | -4.075 | – |

Since D denotes the discriminant score, we can write the following Linear Discriminant Function, in terms of SOFA and APACHE by using the unstandardized coefficients (weights for the variables) given in Table 8.3.

$$D = -4.075 + 0.194 * \text{SOFA} + 0.139 * \text{APACHE}$$

It means the SOFA score will get a weight of 0.194 and APACHE will get 0.139 and the weighted sum of these two scores becomes the value of D after adding the baseline constant of -4.075.

In the present context we have to use unstandardized coefficients only because the presence of a constant is compulsory. The standardized coefficients are also provided by SPSS but they are used only when the D-score (left-hand side of the above model) admits zero as a baseline value (when both SOFA and APACHE are set to zero).

The standardized coefficients shown in Table 8.3 are called the *beta coefficients* in the regression model and represent the relative importance of the independent variable. For instance in this case APACHE is relatively more important than the SOFA in explaining the discrimination between the groups.

Further output from SPSS is understood as follows.

1. There will be two linear functions (called *canonical functions*) developed for the two groups and they are given in the following table.

| Classification Function Coefficients | | |
| --- | --- | --- |
| Variable | Outcome | |
| | 0 | 1 |
| SOFA | 0.652 | 1.102 |
| APACHE | 0.506 | 0.829 |
| (Constant) | -6.884 | -17.8 |

Each function represents a predication formula for the corresponding group.

2. The above two functions are evaluated at the group centroids (means) and the mean scores are shown below.

| Functions at Group Centroids | |
|---|---|
| Outcome | Discriminant score |
| 0 | -0.557 |
| 1 | 1.765 |

The average of these two scores $(1.765-0.557)/2 = 0.604$, is taken as the *cutoff* for classification.

3. Now the decision rule is to classify a new patient into group $= 1$ if the D-score $> 0.604$ (predicting the death) and classify to group $= 0$ otherwise.

4. The actual and predicted membership for each case is produced by SPSS as a table along with several intermediate details. This table is copied into MS-Excel and edited to show only the required values as given in Table 8.4.

5. For instance when SOFA $= 4$ and APACHE $= 12$, simple calculation in the above formula gives D $= -1.6309$. So allot this case to group '0'. It is easy to write a formula in MS-Excel to evaluate this function with the given values of the SOFA and APACHE scores.

6. We may observe that in all the cases where D $> 0.604$, the subject is classified into group '1' and to group '0' otherwise. The cases where the actual and predicted membership did not match are marked by '#' and there are only 6 such misclassified cases.

TABLE 8.4: Actual and predicted groups along with discriminant scores obtained from the model

| S.No | Group | | D-score | S.No | Group | | D-score |
| | Actual | Predicted | | | Actual | Predicted | |
| --- | --- | --- | --- | --- | --- | --- | --- |
| 1 | 0 | 0 | -1.631 | 11 | 1 | 1 | 2.417 |
| 2 | 1 | 1 | 1.946 | 12 | 0 | 0 | -0.633 |
| 3# | 0 | 1 | 1.032 | 13 | 0 | 0 | -1.994 |
| 4 | 0 | 0 | 0.257 | 14 | 0 | 0 | -0.995 |
| 5 | 0 | 0 | -0.051 | 15 | 1 | 1 | 2.139 |
| 6# | 0 | 1 | 1.171 | 16 | 0 | 0 | -0.578 |
| 7 | 0 | 0 | 0.475 | 17 | 1 | 1 | 3.863 |
| 8 | 1 | 1 | 1.503 | 18 | 0 | 0 | -1.328 |
| 9 | 0 | 0 | -1.080 | 19 | 0 | 0 | -0.330 |
| 10 | 0 | 0 | -1.298 | 20# | 0 | 1 | 0.674 |

7. The table of classification shown below indicates the % of correct classifications. It can be seen that 88% of original cases were correctly classified by the proposed discriminant function.

| Actual group | Predicted group | | Total |
| | 0 | 1 | |
| --- | --- | --- | --- |
| 0 | 33 | 5 | 38 |
| 1 | 1 | 11 | 12 |
| Total | 34 | 16 | 50 |

Therefore LDA is a powerful and promising tool to construct new classifiers by combining two or more markers each having its own ability to discriminate between two groups. The new classifier produces a weighted score which we call the *composite score*.

The next aspect of interest is to compare different classifiers and to identify the best. We do this with the help of ROC curve analysis.

## 8.4  ROC Curve Analysis of Discriminant Score

Let us apply ROC curve analysis for the markers SOFA, APACHE and the D-score as a new marker. MedCalc gives the results shown in Figure 8.2.

1. The SOFA score has AUC = 0.912, 95% CI = [0.815 to 0.981], which

means about 91.2% of the cases can be correctly classified with the cutoff SOFA > 10.

2. The APACHE score has AUC = 0.925, 95% CI = [0.798 to 0.974], which means about 92.5% of the cases can be correctly classified with the cutoff APACHE > 17. In terms of AUC this score predicts the event better than the SOFA score.

3. The new D-score proposed from the LDA is a weighted combination of both SOFA and APACHE and when used as a classifier, it has AUC = 0.962, 95% CI = [0.865 to 0.996] and the cutoff for decision making is 0.505. In other words we can predict the outcome of a patient with an accuracy of 96.2% by using the D-score.

Figure 8.2 shows the ROC curve for the D-score of LDA. We also note that the width of the 95% confidence interval for AUC is also small in the case of the D-score which means the chance of correct prediction is more reliable with the D-score than with the other two.

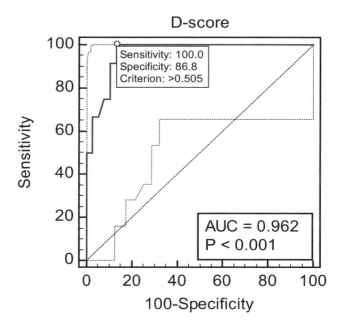

FIGURE 8.2: ROC curves for D-score.

Thus LDA helps in building a model to predict the outcome of an event by combining various predictors (biomarkers) into a single composite score that has better discriminating power than the individual predictors.

LDA can also be extended to situations where the outcome variable has more than two categories (groups). This method is called a *multinomial discriminant* model while with only two groups it is called a *binomial/binary discriminant* model.

In the following section we briefly discuss issues of discrimination among three or more groups using LDA.

---

## 8.5 Extension of Binary Classification

In some applications the researcher encounters situations with more than two populations or categories. For instance, disease staging into 4 groups is one such situation. Classification of anemic patients into a) iron deficiency group, b) B12 deficiency group and c) and both is also a three-group problem.

In such cases the classes are not complimentary to each. In fact when there are k categories we have to distinguish among (k-1) categories using a method like LDA and the last one is automatically identified.

Here are some salient features of this problem.

1. When there are three groups there will be two (3-1) Linear Discriminant functions and each one will be evaluated at the centroids. With k groups there will be (k-1) functions.

2. The table of misclassifications will be different in the case of three categories. For instance let Low, Medium and High be the three categories. A typical classification table shows the following information.

| Actual | Predicted group | | | Total |
|---|---|---|---|---|
| Group | Low | Medium | High | |
| Low | 54* | 10 | 18 | 82 |
| Medium | 34 | 23* | 35 | 92 |
| High | 13 | 23 | 98* | 134 |
| Total | 101 | 56 | 151 | 308 |

\* Indicates correct classification.

3. Out of 308 individuals, the method has classified 175 (56.8%) of cases correctly (58+23+98). The rate of misclassification is 43.2%.

4. Unlike the binary classification, where the decision is based on the cutoff which is the average of the two discriminant functions at the centroid, here we have to use a different measure instead of average of LDF at the

centroids. In fact the classification for a case is based on the Euclidean distance of the new case from the centroids of $k(\neq 3)$ populations. This is a computer-intensive exercise.

5. In a practical situation one has to calculate the Euclidean distance for a new case by writing a simple formula in MS-Excel and decide the group as the one having the smallest Euclidean distance.

Thus LDA is a commonly used tool in machine learning. Several computational tools are available in R and PYTHON to perform LDA.

In the next chapter we discuss another procedure for binary classification, called *logistic regression*.

---

## Summary

Linear Discriminant Analysis (LDA) is a classification procedure in which objects are classified into predefined groups basing on a score obtained from the discriminant function developed from sample data on known characteristics of the individuals. LDA makes use of the information contained in the covariance matrices between groups and within groups. Fisher's discriminant score is developed by running a linear regression model (with several variables) in which the dependent variable Y is dichotomous. The score obtained by the model can be used as a marker which serves as a new classifier. SPSS provides a module to run discriminant analysis. MedCalc offers a handy tool to work out the ROC analysis. It is also possible to carry out the entire analysis using MS-Excel.

---

## Do it yourself (Exercises)

**8.1** Patients suffering from anemia are classified into two groups viz., B12 deficiency group (code = 1) and iron deficiency group (code = 0). Here is a sample data with 30 individuals for whom measurements are taken on 3 parameters X1, X2 and X3 where X1 = Total WBC, X2 = Neutrophils and X3 = Lymphocytes.

| S.No | Group | X1 | X2 | X3 | | S.No | Group | X1 | X2 | X3 |
|---|---|---|---|---|---|---|---|---|---|---|
| 1 | 0 | 8300 | 78 | 18 | | 16 | 1 | 7700 | 45 | 35 |
| 2 | 0 | 5900 | 52 | 39 | | 17 | 1 | 6700 | 76 | 19 |
| 3 | 0 | 5400 | 53 | 39 | | 18 | 1 | 7300 | 70 | 25 |
| 4 | 0 | 9500 | 82 | 12 | | 19 | 1 | 5300 | 61 | 31 |
| 5 | 0 | 6800 | 60 | 29 | | 20 | 1 | 7700 | 49 | 44 |
| 6 | 0 | 7300 | 73 | 21 | | 21 | 1 | 5400 | 64 | 33 |
| 7 | 0 | 5300 | 45 | 45 | | 22 | 1 | 5800 | 54 | 40 |
| 8 | 0 | 4700 | 75 | 17 | | 23 | 1 | 6200 | 66 | 32 |
| 9 | 0 | 6600 | 58 | 28 | | 24 | 1 | 8800 | 73 | 21 |
| 10 | 0 | 7400 | 56 | 23 | | 25 | 1 | 10000 | 76 | 20 |
| 11 | 0 | 6800 | 60 | 29 | | 26 | 1 | 3600 | 38 | 57 |
| 12 | 0 | 4800 | 60 | 26 | | 27 | 1 | 5500 | 64 | 28 |
| 13 | 0 | 9200 | 62 | 34 | | 28 | 1 | 2400 | 55 | 42 |
| 14 | 0 | 12800 | 76 | 18 | | 29 | 1 | 8600 | 75 | 21 |
| 15 | 0 | 2200 | 56 | 41 | | 30 | 1 | 3300 | 54 | 43 |

Perform LDA and obtain the classification formula.

**8.2** The following data refers to a portion of data from a classification problem with class labeled as 0 and 1. There are three predictors X1, X2 and X3 and the data is given below.

| S.No | X1 | X2 | X3 | Class | | S.No | X1 | X2 | X3 | Class |
|---|---|---|---|---|---|---|---|---|---|---|
| 1 | 20 | 1 | 57 | 0 | | 11 | 71 | 0 | 8 | 1 |
| 2 | 19 | 0 | 62 | 0 | | 12 | 26 | 1 | 26 | 1 |
| 3 | 41 | 1 | 74 | 0 | | 13 | 41 | 0 | 14 | 1 |
| 4 | 16 | 0 | 90 | 0 | | 14 | 55 | 0 | 16 | 1 |
| 5 | 45 | 1 | 61 | 0 | | 15 | 22 | 0 | 11 | 1 |
| 6 | 75 | 1 | 80 | 0 | | 16 | 32 | 1 | 6 | 1 |
| 7 | 38 | 1 | 78 | 0 | | 17 | 47 | 1 | 21 | 1 |
| 8 | 40 | 1 | 63 | 0 | | 18 | 56 | 1 | 24 | 1 |
| 9 | 19 | 1 | 87 | 0 | | 19 | 23 | 0 | 12 | 1 |
| 10 | 21 | 0 | 44 | 0 | | 20 | 62 | 0 | 8 | 1 |

Perform Linear Discriminant Analysis to predict the class and derive the rate of misclassification.

**8.3** The following data pertains to the right and left measurements on mandibular and buccalingal bones of 17 female and 25 male individuals observed in a study. The variables are as follows.

| Variable | | Description |
|----------|---|-------------|
| Sex | : | 0=Female, 1=Male |
| M_D_RT (X1) | : | Mandibular right |
| M_D_LT (X2) | : | Mandibular left |
| B_L_RT (X3) | : | Buccalingal right |
| B_L_LT (X4) | : | Buccalingal right |
| IMD (X5) | : | Inter molar distance |

| S.No | Sex | X1 | X2 | X3 | X4 | X5 | | S.No | Sex | X1 | X2 | X3 | X4 | X5 |
|------|-----|----|----|----|----|----|---|------|-----|----|----|----|----|----|
| 1 | 0 | 8.5 | 8.5 | 7.0 | 6.5 | 45.0 | | 22 | 1 | 9.0 | 9.0 | 9.0 | 9.0 | 48.5 |
| 2 | 0 | 8.0 | 8.0 | 6.0 | 6.0 | 46.5 | | 23 | 1 | 9.0 | 9.0 | 7.5 | 7.5 | 51.0 |
| 3 | 0 | 8.0 | 8.0 | 7.0 | 6.5 | 42.0 | | 24 | 1 | 8.0 | 8.0 | 7.0 | 7.0 | 48.0 |
| 4 | 0 | 7.0 | 7.0 | 6.0 | 6.0 | 46.0 | | 25 | 1 | 9.0 | 9.0 | 8.0 | 8.0 | 44.0 |
| 5 | 0 | 7.0 | 7.5 | 6.0 | 6.0 | 45.5 | | 26 | 1 | 8.5 | 8.5 | 8.0 | 8.0 | 50.0 |
| 6 | 0 | 7.0 | 7.0 | 5.0 | 5.0 | 39.0 | | 27 | 1 | 9.0 | 9.0 | 8.5 | 8.5 | 47.0 |
| 7 | 0 | 6.5 | 6.0 | 5.0 | 5.0 | 39.0 | | 28 | 1 | 8.5 | 8.5 | 8.0 | 8.0 | 49.0 |
| 8 | 0 | 7.0 | 7.0 | 5.0 | 5.0 | 38.0 | | 29 | 1 | 8.5 | 8.5 | 7.0 | 7.0 | 49.0 |
| 9 | 0 | 6.5 | 6.5 | 4.5 | 4.5 | 40.0 | | 30 | 1 | 9.0 | 9.0 | 8.0 | 8.0 | 49.0 |
| 10 | 0 | 7.0 | 7.0 | 5.0 | 5.5 | 40.0 | | 31 | 1 | 7.5 | 7.5 | 8.0 | 8.0 | 49.0 |
| 11 | 0 | 6.5 | 6.5 | 5.5 | 5.5 | 38.0 | | 32 | 1 | 7.5 | 8.0 | 7.0 | 8.0 | 45.0 |
| 12 | 0 | 6.5 | 6.0 | 5.5 | 5.0 | 37.0 | | 33 | 1 | 7.0 | 7.5 | 7.0 | 6.5 | 40.5 |
| 13 | 0 | 7.0 | 7.0 | 6.0 | 6.0 | 39.0 | | 34 | 1 | 8.0 | 8.0 | 7.0 | 7.0 | 44.0 |
| 14 | 0 | 7.0 | 6.5 | 5.0 | 5.0 | 39.0 | | 35 | 1 | 8.0 | 8.0 | 8.0 | 8.0 | 43.0 |
| 15 | 0 | 6.0 | 6.0 | 5.5 | 5.5 | 40.0 | | 36 | 1 | 7.0 | 7.0 | 7.0 | 7.0 | 42.0 |
| 16 | 0 | 6.5 | 6.5 | 5.0 | 5.0 | 45.0 | | 37 | 1 | 7.0 | 7.0 | 6.5 | 6.5 | 44.5 |
| 17 | 0 | 6.5 | 6.5 | 5.0 | 5.0 | 39.0 | | 38 | 1 | 7.0 | 7.0 | 7.0 | 7.0 | 42.0 |
| 18 | 1 | 8.5 | 9.0 | 8.0 | 8.5 | 53.0 | | 39 | 1 | 7.5 | 7.5 | 7.0 | 7.0 | 38.0 |
| 19 | 1 | 8.0 | 8.0 | 8.0 | 7.5 | 48.0 | | 40 | 1 | 7.0 | 7.0 | 7.0 | 7.0 | 41.0 |
| 20 | 1 | 9.0 | 9.0 | 8.0 | 8.0 | 51.0 | | 41 | 1 | 7.5 | 8.0 | 7.0 | 7.0 | 40.0 |
| 21 | 1 | 9.0 | 9.0 | 9.0 | 9.0 | 50.0 | | 42 | 1 | 8.0 | 8.0 | 7.0 | 6.5 | 43.0 |

It is desired to predict the gender of the individual basing on the data on the 5 predictors using Linear Discriminant Analysis. How do you predict the gender of a new individual when the data on 5 predictors are given?

8.4 A large international air carrier has collected data on employees in two different job classifications a) customer service personnel and b) mechanics. The director of Human Resources wants to know if these two job classifications appeal to different personality types. Each employee is administered a battery of psychological tests which includes measures of interest in outdoor activity, sociability and conservativeness. A sample of 50 records and the code sheet is given below.

| Variable | | Description |
|---|---|---|
| Job | : | 1 = Customer service and 2 = Mechanic |
| Outdoor activity (X) | : | Score (Continuous variable) |
| Sociability (Y) | : | Score (Continuous variable) |
| Conservativeness (Z) | : | Score (Continuous variable) |

Carry out LDA and obtain a predication formula regarding interest of individual in the type of job.

| S.No | Job | X | Y | Z | S.No | Job | X | Y | Z |
|---|---|---|---|---|---|---|---|---|---|
| 1 | 1 | 10 | 22 | 5 | 26 | 2 | 20 | 27 | 6 |
| 2 | 1 | 14 | 17 | 6 | 27 | 2 | 21 | 15 | 10 |
| 3 | 1 | 19 | 33 | 7 | 28 | 2 | 15 | 27 | 12 |
| 4 | 1 | 14 | 29 | 12 | 29 | 2 | 15 | 29 | 8 |
| 5 | 1 | 14 | 25 | 7 | 30 | 2 | 11 | 25 | 11 |
| 6 | 1 | 20 | 25 | 12 | 31 | 2 | 24 | 9 | 17 |
| 7 | 1 | 6 | 18 | 4 | 32 | 2 | 18 | 21 | 13 |
| 8 | 1 | 13 | 27 | 7 | 33 | 2 | 14 | 18 | 4 |
| 9 | 1 | 18 | 31 | 9 | 34 | 2 | 13 | 22 | 12 |
| 10 | 1 | 16 | 35 | 13 | 35 | 2 | 17 | 21 | 9 |
| 11 | 1 | 17 | 25 | 8 | 36 | 2 | 16 | 28 | 13 |
| 12 | 1 | 10 | 29 | 11 | 37 | 2 | 15 | 22 | 12 |
| 13 | 1 | 17 | 25 | 7 | 38 | 2 | 24 | 20 | 15 |
| 14 | 1 | 10 | 22 | 13 | 39 | 2 | 14 | 19 | 13 |
| 15 | 1 | 10 | 31 | 13 | 40 | 2 | 14 | 28 | 1 |
| 16 | 1 | 18 | 25 | 5 | 41 | 2 | 18 | 17 | 11 |
| 17 | 1 | 0 | 27 | 11 | 42 | 2 | 14 | 24 | 7 |
| 18 | 1 | 10 | 24 | 12 | 43 | 2 | 12 | 16 | 10 |
| 19 | 1 | 15 | 23 | 10 | 44 | 2 | 16 | 21 | 10 |
| 20 | 1 | 8 | 29 | 14 | 45 | 2 | 18 | 19 | 9 |
| 21 | 1 | 6 | 27 | 11 | 46 | 2 | 19 | 26 | 7 |
| 22 | 1 | 10 | 17 | 8 | 47 | 2 | 13 | 20 | 10 |
| 23 | 1 | 1 | 30 | 6 | 48 | 2 | 28 | 16 | 10 |
| 24 | 1 | 14 | 29 | 7 | 49 | 2 | 17 | 19 | 11 |
| 25 | 1 | 13 | 21 | 11 | 50 | 2 | 24 | 14 | 7 |

# Suggested Reading

1. Johnson, R. A., & Wichern, D. W. 2014. *Applied multivariate statistical analysis*, 6th ed. Pearson New International Edition.

2. Kash, K.S. 1982. *Statistical Analysis – An interdisciplinary introduction to Univariate and Multivariate Methods.* New York: Radius Press.

3. Anderson T.W. 2003. *An introduction to Multivariate Statistical Analysis.* $3^{rd}$ edition. New York: John Wiley.

4. Alvin C.Rencher, William F. Christensen. 2012. *Methods of Multivariate Analysis.* $3^{rd}$ ed. Brigham Young University: John Wiley & Sons.

5. Fisher, R.A. 1936. The Use of Multiple Measurement in Taxonomic Problems. *Annals of Eugenics,* 7, 179–188.

# Chapter 9

# Logistic Regression for Binary Classification

9.1    Introduction ..................................................... 169
9.2    Simple Binary Logistic Regression ............................. 170
9.3    Binary Logistic Regression with Multiple Predictors ........... 175
9.4    Assessment of the Model and Relative Effectiveness of
       Markers ....................................................... 179
9.5    Logistic Regression with Interaction Terms ..................... 180
       Summary ....................................................... 181
       Do it yourself (Exercises) ..................................... 182
       Suggested Reading ............................................. 184

In God we trust, all others bring data.

W.E. Deming (1900 – 1993)

## 9.1   Introduction

Logistic Regression (LR) is another approach to handle a binary classification problem. As done in the case of LDA there will be two groups into which the patients were already classified by standard method. We wish to develop a mathematical model to predict the likely group for a new individual.

Let the two groups be denoted by 0 (absence of a condition) and 1 (presence of a condition). Let $P(Y = 1$ given the status) denote the *conditional probability* that a new individual belongs to group 1 given status of the patient in terms of one or more biomarkers. Let $X_1, X_2, ..., X_k$ be k explanatory variables (markers) which may include qualitative variables like gender or disease status measured on a nominal or ordinal scale along with continuous variables. In LDA we have assumed that the data on the explanatory variables is continuous and follow normal distribution. In contrast, the LR approach does not make any assumptions and hence is considered to be more robust than the LDA.

The simplest form of the LR model is the one involving a single predictor and a response variable (dichotomous or multinomial). When there are two or more predictors and a binary outcome, we call it *binary multiple logistic regression*. When the outcome contains two or more levels like 0, 1, 2, ..., m, we call it *multinomial logistic regression*.

## 9.2   Simple Binary Logistic Regression

Let Y denote the binary outcome with values 0 and 1 indicating the absence and presence of a condition.

The binary logistic regression model estimates the probability that the event of interest occurs when X value is given. It means we estimate P (Y=1 given X). Thus the outcome variable is a probability whose value lies between 0 and 1. The model is given by

$$P(Y = 1 \mid X = x) = \frac{e^{(\beta_0 + \beta_1 x)}}{1 + e^{(\beta_0 + \beta_1 x)}} \tag{9.1}$$

The quantity $\{\beta_0 + \beta_1 x\}$ represents the outcome from a linear regression model with intercept $\beta_0$ and slope (regression coefficient) $\beta_1$. This value is exponentiated and divided by the same quantity plus 1. As a result the left-hand side of Equation 9.1 will always be a number between 0 and 1 and hence represents the proportion of cases with label $= 1$ when the input is x.

For instance let $\beta_0 = 0$ and $\beta_1 = 1$. For different values of x, the curve of $P(Y = 1)$ against x is shown in Figure 9.1, which is called a *logistic curve*.

The logistic function given in Equation 9.1 is usually written as

$$p = \frac{e^{(\beta_0 + \beta_1 x)}}{1 + e^{(\beta_0 + \beta_1 x)}} \tag{9.2}$$

where p denotes $P(Y = 1)$ which is always understood as probability of the desired outcome.

Taking a natural logarithm on both sides, Equation 9.2 reduces to

$$\ln\left(\frac{p}{1-p}\right) = \beta_0 + \beta_1 x \tag{9.3}$$

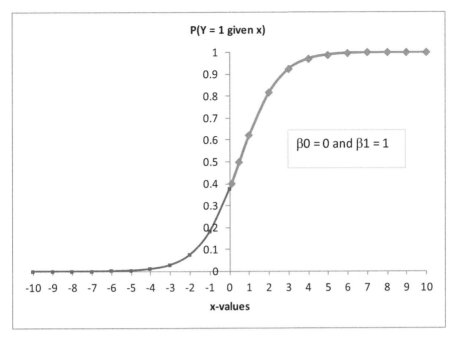

FIGURE 9.1: Behaviour of the logistic curve.

The interesting feature of this curve is that it takes values only between 0 and 1 irrespective of the values of x. When $x = 0$ we get $P(Y = 1) = 0.5$ since we have taken $\beta_0 = 0$ and $\beta_1 = 1$. It means for an individual having a value $x = 0$, there is 50% chance of being categorized into group-1.

Let $y_i$ denote the binary outcome (0 or 1) of the $i^{th}$ individual for whom $x_i$ (predictor value) is given. Then,

a) $y_i$ follows binomial distribution with proportion (parameter), p.

b) The mean value of $y_i$ is given by the logistic function given in Equation 9.2.

In other words we wish to estimate the "mean of y given x" from the model given in Equation 9.2 or its transformation given in Equation 9.3. The parameters $\beta_0$ and $\beta_1$ are found by maximum likelihood method and computer software can be used to do this.

The data for estimation requires $n_1$ records from group-0 and $n_2$ from group-1 where the outcome is already known. Once the coefficients are estimated as $b_0$ and $b_1$, we can estimate (predict) p from Equation 9.2 for a given value of x.

Once the regression coefficients $b_0$ and $b_1$ are found from the data, a simple substitution of x value in Equation 9.3 gives the predicted value of Y, denoted by Y-cap, as a continuous number (the original Y is 0 or 1). This number is converted into probability p, which is a number between 0 and 1 by using Equation 9.2.

For instance let $b_0 = 1.258$ and $b_1 = 0.56$ be the estimated coefficients. If we take x = 1.30 we get $\ln\left(\dfrac{p}{1-p}\right) = 1.986$ from Equation 9.3 so that p = 0.8793. It means the new case with x = 1.30 has about 87.93% chance to have come from the group for which Y = 1.

Suppose we take p = 0.50 for a new case. There is a 50% chance of being allotted to group-0 or group-1 and the situation is indecisive. So one rule of classification is as follows:

*"Allot a new case to group-1 if p > 0.50 and to group-0 otherwise"*

At the end of classification we may arrive at few misclassifications because the allotment to groups is based on a chance mechanism. As discussed in the discriminant analysis in Chapter 8, we can prepare a table of misclassification and find the % of correct classifications.

Consider the following illustration.

**Illustration 9.1** Reconsider the ICU scores data used in Illustration 7.1. A portion of data with the 20 patients, who were classified as alive or dead, with their corresponding APACHE score is given in Table 9.1. The analysis is however done on 50 records of the original data.

It is desired to examine whether APACHE score can be used as a predictive marker for the death of a patient. What is the cutoff score? What would be the percentage of correct classifications?

TABLE 9.1: ICU scores data with APACHE and outcome

| S.No | Outcome | APACHE | S.No | Outcome | APACHE |
|------|---------|--------|------|---------|--------|
| 1  | 0 | 12 | 11 | 1 | 23 |
| 2  | 1 | 21 | 12 | 0 | 15 |
| 3  | 0 | 20 | 13 | 0 | 8  |
| 4  | 0 | 20 | 14 | 0 | 11 |
| 5  | 0 | 15 | 15 | 1 | 21 |
| 6  | 0 | 21 | 16 | 0 | 14 |
| 7  | 0 | 16 | 17 | 1 | 32 |
| 8  | 1 | 22 | 18 | 0 | 10 |
| 9  | 0 | 9  | 19 | 0 | 13 |
| 10 | 0 | 13 | 20 | 0 | 23 |

**Analysis:** We can use SPSS to run a logistic regression model with X = APACHE score and Y = Outcome.

By using SPSS options Analyse → Regression → Binary Logistic we get the following linear regression model. (Details with multiple independent variables are explained in the following section)

$$\ln\left(\frac{p}{1-p}\right) = -9.553 + 0.444 * \text{APACHE} \tag{9.4}$$

This model has a measure of goodness of fit given by Nagelkerke $R^2 = 0.609$ which means about 60% of $\ln\left(\frac{p}{1-p}\right)$ is explained by this model given in Equation 9.4.

Now for each case of the data when the APACHE score is substituted in Equation 9.4 and transforming the resulting values to probabilities using Equation 9.2, we get the probability of group membership. All this is automatically done by SPSS by choosing the Save option and it in turn creates two additional columns in the data file. These columns are filled with the following entries for each record.

a) *Probabilities:* This is the predicted probability of group membership (a number between 0 and 1)

b) *Group membership:* This gives the membership of each case, as predicted by the model. When the predicted probability is greater than 0.50 the case will be predicted as 1, else as 0.

The classification results are shown in Table 9.2.

TABLE 9.2: Classification of patients using APACHE as predictor

| S.No | Group | | Pred. Prob. | S.No | Group | | Pred. Prob. |
|---|---|---|---|---|---|---|---|
| | Actual | Predicted | | | Actual | Predicted | |
| 1 | 0 | 0 | 0.014 | 11 | 1 | 1 | 0.658 |
| 2 | 1 | 0 | 0.442 | 12 | 0 | 0 | 0.052 |
| 3 | 0 | 0 | 0.337 | 13 | 0 | 0 | 0.003 |
| 4 | 0 | 0 | 0.337 | 14 | 0 | 0 | 0.009 |
| 5 | 0 | 0 | 0.052 | 15 | 1 | 0 | 0.442 |
| 6 | 0 | 0 | 0.442 | 16 | 0 | 0 | 0.034 |
| 7 | 0 | 0 | 0.079 | 17 | 1 | 1 | 0.991 |
| 8 | 1 | 1 | 0.553 | 18 | 0 | 0 | 0.006 |
| 9 | 0 | 0 | 0.004 | 19 | 0 | 0 | 0.022 |
| 10 | 0 | 0 | 0.022 | 20 | 0 | 1 | 0.658 |

Pred. Prob.= Predicted Probability.

From Table 9.2, we see that the method has produced 8 wrong classifications amounting to 84% of correct classification. The table of classifications is shown below.

| Observed | Predicted* | | % Correctly Classified |
|---|---|---|---|
| | Alive | Dead | |
| Alive | 35 | 3 | 92.1 |
| Dead | 5 | 7 | 58.3 |
| Overall % of correct classification = 84.0 | | | |

\* Cut value is 0.500 for the discriminant score.

With this classification, 3 out of 38 cases which are alive were misclassified as dead and 5 out of 12 dead cases were classified as alive. The percentage of correct classification is a measure of the efficiency of the fitted model for predicting the outcome. A perfect model is expected to have zero misclassifications, which cannot happen in practice. There could be several reasons for this misclassification.

a) The data is not sufficient to estimate the coefficients accurately.

b) There may be other predictor variables, not included in the model. For instance factors like shock, history of diabetes, hypertension etc., might have been included.

c) The linear relationship chosen in the model may not be a good fit. For instance if the R-square value was very low, like 0.11, we have to check the model before making a prediction.

For the above problem instead of logistic regression, suppose we have used Linear Discriminant Analysis discussed in Chapter 8 with only APACHE. We get the same result of 84% correct classification but the number of misclassifications is different as shown below.

| Observed | Predicted* | | % Correctly Classified |
|---|---|---|---|
| | Alive | Dead | |
| Alive | 31 | 7 | 81.58 |
| Dead | 1 | 11 | 91.67 |
| Overall % of correct classification = 84.0 | | | |

\* Cut value is 1.044 for the discriminant score.

The difference is basically due to the approach used for classification. Assessing the merits and demerits of the two approaches for classification is beyond the scope of this book. However, logistic regression has fewer assumptions about the data than the LDA and hence it can be used as a predictive model.

In the following section we discuss a binary logistic model with multiple predictors.

---

## 9.3 Binary Logistic Regression with Multiple Predictors

We can also accommodate more than one independent variable in developing a binary logistic regression. To do this, we first develop a multiple linear regression model, convert it into a logistic model and determine the probability of classification.

SPSS has the following options to run the logistic regression.

1. We can select all the variables that are likely to influence the outcome.

2. We can include both continuous and categorical variables into the model.

3. We can also handle interaction terms in estimating the model.

4. By choosing the option *forward conditional*, we can progressively include only a few variables that contribute significantly to the model.

5. We can choose the option to save the predicted score and the group membership to the data file.

6. In the case of categorical variables like gender, we should specify the reference category as 'first' or 'last' in the list. For instance, the variable 'Ventilator' with codes 1 or 0 means 'Required' and 'Not required' respectively. The reference category could be taken as '0'. For this type of variable we calculate a new measure called *odds ratio* which expresses the magnitude and direction of the effect of the categorical factor on the outcome.

7. The procedure ultimately produces the stepwise results indicating how the $R^2$ value of the regression model improves at each step and which variables are included into the model. All the variables that are excluded from the analysis will also be displayed in the output.

8. The table of classification as predicted by the model and the percentage of correct classification will also be displayed.

Here is an illustration of a logistic regression with multiple predictors.

**Illustration 9.2** Reconsider the ICU scores data used in Illustration 7.1. A portion of the data with 15 records is shown in Table 9.2 with selected variables, to illustrate the method of handling categorical variables in logistic regression along with continuous variables. Analysis is however carried out on the first 50 records of the original data. It is desired to predict the end event, outcome=1 (death) given the inputs observed on the patient.

TABLE 9.3: ICU scores data with selected variables

| S.No | Age | Gen der | Diag nosis | SOFA | APACHE | AKI | Shock | Out come | Venti lator |
|------|-----|---------|------------|------|--------|-----|-------|----------|-------------|
| 1 | 20 | 0 | 3 | 4 | 12 | 0 | 1 | 0 | 1 |
| 2 | 52 | 1 | 2 | 16 | 21 | 1 | 0 | 1 | 1 |
| 3 | 25 | 1 | 2 | 12 | 20 | 1 | 0 | 0 | 0 |
| 4 | 53 | 1 | 2 | 8 | 20 | 0 | 1 | 0 | 1 |
| 5 | 15 | 1 | 2 | 10 | 15 | 0 | 1 | 0 | 1 |
| 6 | 40 | 1 | 3 | 12 | 21 | 1 | 1 | 0 | 1 |
| 7 | 70 | 1 | 2 | 12 | 16 | 1 | 0 | 0 | 1 |
| 8 | 50 | 1 | 2 | 13 | 22 | 1 | 1 | 1 | 1 |
| 9 | 27 | 1 | 3 | 9 | 9 | 0 | 1 | 0 | 0 |
| 10 | 30 | 0 | 2 | 5 | 13 | 0 | 1 | 0 | 1 |
| 11 | 47 | 1 | 3 | 17 | 23 | 1 | 1 | 1 | 1 |
| 12 | 23 | 0 | 2 | 7 | 15 | 0 | 0 | 0 | 1 |
| 13 | 19 | 0 | 2 | 5 | 8 | 0 | 0 | 0 | 0 |
| 14 | 40 | 0 | 2 | 8 | 11 | 0 | 0 | 0 | 1 |
| 15 | 30 | 0 | 3 | 17 | 21 | 1 | 1 | 1 | 1 |

**Analysis:**

Here the variable outcome is the dependent variable which is binary. There are 8 predictors out of which three are measurements and the others are categorical. We proceed with the SPSS options by fixing the 'Dependent' variable as the outcome and all the 10 predictors as 'Covariates' as shown in Figure 9.2. The method of selection of variables will be *forward conditional*.

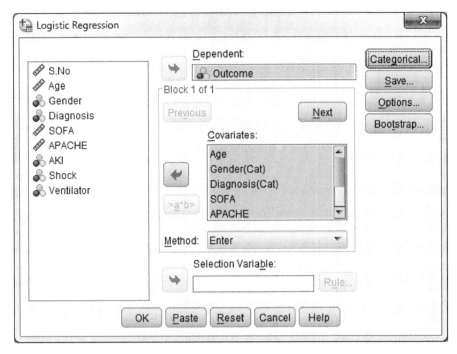

FIGURE 9.2: Selection of continues and categorical variables.

Clicking on the tab 'Categorical', all covariates except age, SOFA and APACHE go into the 'Categorical Covariates' list. Press 'Continue' and press 'OK'.

The categorical variables are defined as indicators with last category as reference. The inclusion of a variable into the model and the removal of a variable from the model are taken as per the default probabilities.

When the model is run the following output is produced.

**Model summary:**

It shows how the regression was built up in 3 steps and at each step a measure of goodness is also given. We can use Nagelkerke R Square as the measure and we find that at the 3rd iteration, the procedure is stopped with $R^2 = 0.808$.

| Step | -2Log likelihood | Cox & Snell R Square | Nagelkerke R Square |
|------|------------------|----------------------|---------------------|
| 1 | 28.995[a] | 0.407 | 0.609 |
| 2 | 21.700[a] | 0.487 | 0.730 |
| 3 | 16.357[b] | 0.539 | 0.808 |

a. Estimation terminated at iteration number 7 because parameter estimates changed by less than 0.001.

b. Estimation terminated at iteration number 9 because parameter estimates changed by less than 0.001.

**Variables in the equation:**

The variables that were selected into the model in successive steps are shown here. We will view only the last step since it contains the final model. The details are shown below.

| Source | B | S.E. | Wald | df | Sig. | Exp(B) | 95% C.I. |
|--------|-----|------|------|-----|------|--------|----------|
| SOFA | 1.210 | 0.585 | 4.281 | 1 | 0.039 | 3.354 | [ 1.066, 10.557 ] |
| APACHE | 0.703 | 0.339 | 4.313 | 1 | 0.038 | 2.020 | [ 1.04, 3.922 ] |
| AKI(1) | 4.044 | 2.184 | 3.428 | 1 | 0.064 | 57.048 | [ 0.789, 4125.881 ] |
| Constant | -30.097 | 13.441 | 5.014 | 1 | 0.025 | 0.000 | |

The above table shows for each predictor, the regression coefficient (B), its standard error (SE) and the p-value (denoted by Sig. basing on the t-test with null hypothesis of zero coefficients). It also shows the odds ratio, denoted by Exp(B)or $e^B$ and its 95% CI.

From the above table we can write the model as

$$\ln\left(\frac{p}{1-p}\right) = -30.097 + 1.210 * SOFA + 0.703 * APACHE + 4.044 * AKI$$

Suppose for a patient the values of the SOFA and APACHE scores, are only known and the patient is present with AKI. Then substituting the SOFA and APACHE scores and putting $AKI = 1$ in the above equation we get a value of $\ln\left(\frac{p}{1-p}\right)$, and $\ln\left(\frac{p}{1-p}\right) > 0$ predicts death, which means $p > 0.50$.

This exercise is repeated for all 50 patients in the data set and the predicted group probability (p) and the corresponding group membership, as predicted by the model will be saved in the data file. Both the predicted group probability and predicted group membership are saved to the data file so that the misclassifications, if any, can be noticed from the data file, casewise.

**The Odds Ratio (OR):**

Logistic regression helps in predicting the outcome in the presence of both continuous and categorical variables. In this case AKI is a categorical variable for which the OR = 57.048 which means that those presented with AKI will have 57 times more risk of death compared to those without AKI. For every categorical variable that is included in the model we have to interpret the OR.

Sometimes we get very high odds ratios running into several thousands. It is difficult to explain them properly. This usually happens when a classification table contains very small numbers near to zero. Advanced programs written in the R-language have some solutions to overcome this difficulty (Penalized Maximum Likelihood Estimation).

**Classification table:**

This table is by default presented for each step but we show the final step only.

| Observed | Predicted* | | % Correctly Classified |
|----------|-------|------|------------|
| | Alive | Dead | |
| Alive | 37 | 1 | 97.4 |
| Dead | 3 | 9 | 75.0 |
| Overall % of correct classification = 92.0 | | | |

\* Cut value is 0.500 for the discriminant score.

It is easy to see that this logistic regression model has misclassified only 4 out of 50 cases which amounts to 92% of accuracy.

In the following section we address the issues in assessing the predictive ability of biomarkers using the LR model.

---

## 9.4 Assessment of the Model and Relative Effectiveness of Markers

We can perform ROC analysis (discussed in Chapter 7) to compare the diagnostic accuracy of several individual variables (markers) and pick up the one having the highest AUC. Instead of a single variable as a marker, the score obtained from the LR model itself serves as a composite marker and we can find the corresponding AUC. In the above example if we perform ROC curve analysis we get the results shown below.

| Marker | AUC | Std. Error | p-value | 95% CI |
|--------|-----|------------|---------|--------|
| AKI | 0.607 | 0.039 | 0.0071 | [ 0.531, 0.683 ] |
| SOFA | 0.836 | 0.028 | 0.0001 | [ 0.780, 0.891 ] |
| APACHE | 0.920 | 0.020 | 0.0001 | [ 0.880, 0.959 ] |
| LR Model (Score) | 0.920 | 0.018 | 0.0001 | [ 0.884, 0.956 ] |

It can be seen that the AUC is the same for the both the APACHE score and the score produced by the LR model. It means the APACHE alone can predict the end event as well as that of the LR model score. However, the LR model predicts the event with a lower standard error, which means it gives a more consistent (stable) performance than the APACHE score alone.

## 9.5   Logistic Regression with Interaction Terms

Sometimes two or more factors and their joint effect will influence the outcome. Let $X_1$ and $X_2$ be two predictors of the binary outcome Y. Then the joint effect of $X_1$ and $X_2$ is represented in the model as

$$\ln\left\{\frac{p}{1-p}\right\} = \beta_0 + \beta_1 X_1 + \beta_2 X_2 + \beta_3(X_1 * X_2) \tag{9.5}$$

The estimation of coefficients is similar to what was done in the multiple linear regression with interaction (Chapter 6)

Here is an illustration.

**Illustration 9.3** To observe the effect with the interaction term, reconsider the data given in Illustration 9.2. Here the response variable is the Outcome (Y) and the predictors APACHE $(X_1)$, SOFA $(X_2)$ and APACHE * SOFA $(X_1 * X_2)$.

The predictors $X_1$ and $X_2$ are usually considered as main effects and $(X_1 * X_2)$ is the interaction effect.

The SPSS instructions are given below.

a) Send the Outcome variable to the 'Dependent' tab and 'APACHE, SOFA' to the Covariates pane.

b) To create the interaction term, select APACHE and SOFA simultaneously using 'ctrl' key and click button having '>a*b>'.

c) Click 'options' tab and choose 'Hosmer-Lemeshow goodness of fit' and 'CI of exp(B)' with 95% and press Continue.

d) Finally press OK to view the results.

The following observations are made from the output.

**Model Summary:**

The Nagelkarke $R^2$ is 0.643. It means about 64.3% of the individuals' status can be predicted using the information on the predictors.

**Classification Table:** The model with interaction terms yields the following table of classification.

| Observed | Predicted* | | % Correctly Classified |
|---|---|---|---|
| | Alive | Dead | |
| Alive | 57 | 21 | 73.1 |
| Dead | 12 | 158 | 92.9 |
| Overall % of correct classification = 86.7 | | | |

\* Cut value is 0.500 for the discriminant score.

The percentage of correct classification is 86.7. Let us recall that the percentage of correct classification obtained without interaction term was only 84.6.

This shows that adding the interaction term helps in improving the model performance in minimizing the misclassification rate.

**Final model:**

Using the 'B' coefficients in the output, the final equation for the LR model with interaction will take the form

$$\ln\left(\frac{p}{1-p}\right) = 3.974 + 0.257 * \text{SOFA} - 0.142 * \text{APACHE}$$
$$-0.018 * (\text{APACHE} * \text{SOFA})$$

The classification is done as usual by substituting the inputs.

We end this chapter with the observation that Logistic Regression is an effective method for binary classification. The computational tool has gained prime importance in clinical research, public health, business, psychology, education and several such fields. This is also a popular tool in machine learning.

## Summary

Binary Logistic regression is a popular tool for binary classification. Unlike the LDA this method does not assume normality of the variables under study. Further, we can handle both continuous and categorical variables in logistic regression and estimate the odds ratio. Predictive models for several events in clinical studies can be developed by using the LR model. In addition to SPSS, several statistical software programs like R, STATA, SAS and even MS-Excel (with Real Statistics Add-ins) and MedCalc provide modules to perform this analysis. When the end event of interest is not binary but has 3 or more options, we have to use a multinomial logistic regression and SPSS has a tool to work with it also.

## Do it yourself (Exercises)

**9.1** Reconsider the data in Exercise 8.2 and perform binary logistic regression analysis.

**9.2** Silva,J.E., Marques de Sá, J.P., Jossinet, J. (2000) carried out a study on the classification of breast tissues and breast cancer detection. A portion of the data used by them is given below. The description of variables and complete data can be found at (http://archive.ics.uci.edu/ml/datasets/breast+tissue). The following data refers to the features of breast cancer tissue measured with eight variables on 106 patients. The tissues were classified as non-fatty tissue (code=1) and fatty tissue (code=0).

| S.No | I0 500 | PA | HFS | DA | Area | A/DA | Max IP | DR | P | Class |
|------|--------|------|-------|--------|----------|-------|--------|--------|---------|-------|
| 1 | 524.79 | 0.19 | 0.03 | 228.80 | 6843.60 | 29.91 | 60.21 | 220.74 | 556.83 | 1 |
| 2 | 330.00 | 0.23 | 0.27 | 121.15 | 3163.24 | 26.11 | 69.72 | 99.09 | 400.23 | 1 |
| 3 | 551.88 | 0.23 | 0.06 | 264.81 | 11888.39 | 44.90 | 77.79 | 253.79 | 656.77 | 1 |
| 4 | 380.00 | 0.24 | 0.29 | 137.64 | 5402.17 | 39.25 | 88.76 | 105.20 | 493.70 | 1 |
| 5 | 362.83 | 0.20 | 0.24 | 124.91 | 3290.46 | 26.34 | 69.39 | 103.87 | 424.80 | 1 |
| 6 | 389.87 | 0.15 | 0.10 | 118.63 | 2475.56 | 20.87 | 49.76 | 107.69 | 429.39 | 1 |
| 7 | 290.46 | 0.14 | 0.05 | 74.64 | 1189.55 | 15.94 | 35.70 | 65.54 | 330.27 | 1 |
| 8 | 275.68 | 0.15 | 0.19 | 91.53 | 1756.24 | 19.19 | 39.31 | 82.66 | 331.59 | 1 |
| 9 | 470.00 | 0.21 | 0.23 | 184.59 | 8185.36 | 44.34 | 84.48 | 164.12 | 603.32 | 1 |
| 10 | 423.00 | 0.22 | 0.26 | 172.37 | 6108.11 | 35.44 | 79.06 | 153.17 | 558.28 | 1 |
| 11 | 1724.09 | 0.05 | -0.02 | 404.13 | 3053.97 | 7.56 | 71.43 | 399.19 | 1489.39 | 0 |
| 12 | 1385.67 | 0.09 | 0.09 | 202.48 | 8785.03 | 43.39 | 143.09 | 143.26 | 1524.61 | 0 |
| 13 | 1084.25 | 0.07 | 0.00 | 191.90 | 2937.97 | 15.31 | 66.56 | 179.98 | 1064.10 | 0 |
| 14 | 649.37 | 0.11 | 0.02 | 207.11 | 3344.43 | 16.15 | 50.55 | 200.85 | 623.91 | 0 |
| 15 | 1500.00 | 0.06 | 0.05 | 375.10 | 4759.46 | 12.69 | 78.45 | 366.80 | 1336.16 | 0 |
| 16 | 770.00 | 0.04 | 0.00 | 175.02 | 346.09 | 1.98 | 25.22 | 173.19 | 654.80 | 0 |
| 17 | 650.00 | 0.04 | 0.15 | 216.81 | 427.53 | 1.97 | 33.77 | 214.17 | 528.70 | 0 |
| 18 | 691.97 | 0.03 | 0.09 | 190.68 | 304.27 | 1.60 | 23.98 | 189.16 | 594.32 | 0 |
| 19 | 1461.75 | 0.04 | 0.05 | 391.85 | 5574.00 | 14.23 | 57.23 | 387.64 | 1428.84 | 0 |
| 20 | 1496.74 | 0.10 | 0.08 | 640.28 | 11072.00 | 17.29 | 108.29 | 631.05 | 1178.27 | 0 |

* Only two decimals displayed in this table.

Perform LR and obtain the model. Determine the cutoff and find the percentage of misclassification? Use all the variables in deriving the model.

**9.3** Reconsider the data in Exercise 8.1, and obtain the LR model by choosing any one variable selection method.

**9.4** A Study on classifying the status of Acute Lymphoid Leukemia (ALL) as 'Alive' or 'Dead' is carried out and a sample of 24 records with Age (in years), Sex (0=Female; 1= Male), Risk (1=High; 2=Standard), Duration of Symptoms (in weeks), Antibiotic Before Induction (1=Yes, 0=No), Platelets, Creatinine, Albumin and Outcome (1=Dead; 0=Alive) are given in the Table 9.4.

TABLE 9.4: Acute Lymphoid Leukemia Data

| S.No | Age | Sex | Risk | PS* | Dur# | Abx** | Platelets | Creatinine | Albumin | Outcome |
|------|-----|-----|------|-----|------|-------|-----------|------------|---------|---------|
| 1 | 41 | 0 | 1 | 1 | 12.00 | 1 | 110000 | 1.4 | 4.5 | 0 |
| 2 | 18 | 0 | 1 | 1 | 9.00 | 0 | 26000 | 0.7 | 3.5 | 0 |
| 3 | 14 | 0 | 1 | 1 | 4.00 | 0 | 19000 | 0.6 | 3.2 | 0 |
| 4 | 4 | 1 | 2 | 4 | 0.57 | 1 | 50000 | 0.7 | 2.8 | 0 |
| 5 | 45 | 0 | 1 | 2 | 4.00 | 0 | 20000 | 0.7 | 3.2 | 1 |
| 6 | 10 | 0 | 1 | 2 | 1.00 | 1 | 59000 | 0.6 | 3.7 | 0 |
| 7 | 7 | 0 | 2 | 1 | 24.00 | 0 | 350000 | 0.6 | 4.1 | 0 |
| 8 | 18 | 0 | 1 | 1 | 8.00 | 1 | 43000 | 1.5 | 3.3 | 1 |
| 9 | 2 | 0 | 1 | 1 | 2.00 | 0 | 101000 | 0.7 | 2.6 | 0 |
| 10 | 5 | 0 | 2 | 1 | 8.00 | 0 | 162000 | 0.5 | 3.6 | 1 |
| 11 | 23 | 1 | 1 | 2 | 1.50 | 1 | 38000 | 0.8 | 2.7 | 0 |
| 12 | 16 | 0 | 1 | 2 | 1.00 | 0 | 45000 | 1.0 | 3.9 | 0 |
| 13 | 22 | 1 | 1 | 2 | 4.00 | 1 | 28000 | 1.0 | 1.8 | 0 |
| 14 | 30 | 1 | 1 | 1 | 4.00 | 0 | 22000 | 1.9 | 3.5 | 0 |
| 15 | 4 | 1 | 2 | 1 | 1.50 | 0 | 30000 | 1.0 | 4.0 | 0 |
| 16 | 16 | 1 | 1 | 1 | 24.00 | 1 | 80000 | 0.8 | 3.8 | 0 |
| 17 | 20 | 0 | 1 | 1 | 4.00 | 0 | 15000 | 0.9 | 2.8 | 0 |
| 18 | 8 | 1 | 1 | 1 | 1.50 | 1 | 11000 | 0.6 | 3.7 | 0 |
| 19 | 1.5 | 0 | 1 | 1 | 1.50 | 1 | 64000 | 0.6 | 3.9 | 1 |
| 20 | 16 | 0 | 2 | 1 | 3.00 | 0 | 348000 | 1.4 | 3.6 | 0 |
| 21 | 4 | 0 | 2 | 1 | 12.00 | 0 | 49000 | 0.7 | 4.5 | 0 |
| 22 | 36 | 0 | 1 | 4 | 4.00 | 0 | 582000 | 0.9 | 2.7 | 0 |
| 23 | 17 | 1 | 1 | 2 | 2.00 | 0 | 74000 | 0.7 | 3.8 | 1 |
| 24 | 3 | 0 | 2 | 2 | 2.00 | 0 | 45000 | 0.7 | 3.4 | 0 |

*(Data courtesy: Dr. Biswajit Dubashi and Dr. Smita Kayal, Department of Medical Oncology, JIPMER, Puducherry.)*
*PS: Performance Status; #Dur: Duration of Symptoms; **Abx: Antibiotics Before Induction.

Perform the LR analysis and comment on the following a) Nagelkarke $R^2$ b) LR Model c) exp(B) and d) classification table.

---

# Suggested Reading

1. Johnson, R. A., & Wichern, D. W. 2014. *Applied multivariate statistical analysis*, 6th ed. Pearson New International Edition.

2. Anderson T.W. 2003, An introduction to Multivariate Statistical Analysis, 3rd edition, John Wiley, New York.

3. Silva,J.E., Marques de Sá, J.P., Jossinet, J.(2000). UCI Machine Learning Repository [http://archive.ics.uci.edu/ml]. Irvine, CA: University of California, School of Information and Computer Science.

# Chapter 10

## Survival Analysis and Cox Regression

10.1    Introduction ..................................................... 185
10.2    Data Requirements for Survival Analysis ....................... 186
10.3    Estimation of Survival Time with Complete Data
        (No Censoring) ................................................ 187
10.4    The Kaplan–Meier Method for Censored Data ................. 190
10.5    Cox Regression Model ........................................ 193
        Summary ..................................................... 199
        Do it yourself (Exercises) .................................. 199
        Suggested Reading ........................................... 203

"All models are wrong, but some are useful."

George E. P. Box (1919 – 2013)

## 10.1    Introduction

Survival analysis is a tool that helps in estimating the *chance of survival* of an individual (patient or equipment) before reaching a critical event like death or failure. Also known as *risk estimation*, survival analysis is a key aspect in health care particularly with chronic diseases. The objective is to predict the *residual life* after the onset/management of a bad health condition. The outcome of such studies will be binary indicating death or survival at a given time point. This applies even for equipment which faces *failures* and needs intervention eg., repair or maintenance.

We generally use the word *hazard* to indicate an unfavorable state such as disturbance, failure or death and the complimentary situation is *hazard-free* state also known as *survival state*. Every individual will be in one of these states but not both at a given point in time. When a failure (complaint) occurs it is sometimes possible to repair (intervention) and return to normal or a failure-free state. Recurrence of such events and recovery to a normal state is often measured in terms of *time between failures* and one simple measure is the *Mean Time Between Failures* (MTBF).

In clinical management of patients, an event like death is non-repairable once it occurs. All interventions are valid only before the occurrence of the event. In such cases we measure the time to occurrence and a simple measure is the Mean Time To Failure (MTTF) which sounds like a warranty period.

Survival time is the time elapsed between the onset of a condition till the occurrence of death (failure) of the patient or machine. Two important objectives arise in this type of study.

a) To estimate the time of the event of interest (death or failure).

b) To identify the factors or covariates that are related to the outcome.

In the following section we discuss briefly the type of data required to perform survival analysis.

---

## 10.2   Data Requirements for Survival Analysis

The data required for survival analysis is known as *time-to-event* data and should contain the following entities.

1. An event of interest indicating the binary outcome: alive (0) or dead (1), relapse or no-relapse etc.

2. A continuous variable measuring the 'time-to-event' (e.g., time to relapse or time to discharge from ICU).

3. One or more prognostic factors (categorical or continuous) that may influence the time to event.

In a practical context, the survival times are recorded for each subject for a specified *follow-up period* say 6 or 12 months in some cases and several years in long-term clinical trials. For all the patients recruited in the study, the survival is recorded at every point of observation (daily, weekly or monthly) and if the event of interest does not happen, the survival time increases by another unit of time since the last visit.

Interestingly complete data on a patient may not be available for the entire follow-up period for the following reasons.

a) The event of interest did not happen.

b) The patient does not visit for a check-up and status is unknown.

c) Follow-up stopped since there is enough evidence to meet the research objectives.

As a result of this, there will be incomplete data on the survival time of patients during the study period. Such cases are said to be *censored*. For the rest of the study period the survival is unknown. We cannot also classify the case as non-surviving. As such the censored cases create *incomplete data* in the time domain.

For the above reason, the classical statistical methods like mean or median of survival time will not be valid estimates of the true survival time. Using conditional probabilities, a method of estimating the survival time was proposed by Kaplan and Meier (1958) which is known as *product-limit estimate*.

The type of censoring used in survival analysis is often known as *right censoring* to mean that there is an upper limit beyond which observations are not made.

Before using the methods of survival analysis let us understand how the survival probabilities are calculated from clinical data. Here are some basic concepts related to survival analysis.

---

## 10.3 Estimation of Survival Time with Complete Data (No Censoring)

Suppose there are n individuals and all of them were followed up until death. It means the complete days of survival of each individual are available. Let $t_1, t_2, \ldots, t_n$ be the times at which the patients died. We can sort these times in an ascending order and label them as $t_{(1)}, t_{(2)}, \ldots, t_{(n)}$ where the bracket in the subscript indicates ordered data. Then for the $i^{th}$ patient, the survival function is denoted by $S(t_{(i)}) = P(\text{the individual survives longer than } t_{(i)})$ where $P(.)$ denotes the probability.

From sample data this function is estimated as

$$\widehat{S}(t_{(i)}) = \frac{(n-i)}{n} \; \forall \, i = 1, 2, \ldots, n \tag{10.1}$$

There will be (n-i) individuals surviving at the observation time $t_{(i)}$ and hence the survival function is simply the proportion of individuals surviving (out of n). The value produced by (Equation 10.1) can be interpreted as the probability of survival of the $i^{th}$ individual up to time t. It also follows that $\widehat{S}(t_{(0)}) = 1$ and $\widehat{S}(t_{(}(n)) = 0$ because at the start of time all individuals are surviving and at the end no one is found surviving.

While computing $\widehat{S}(t_{(i)})$ in Equation 10.1 suppose there are tied values, say 4 individuals had the same survival time. Then the largest 'i' value will be used for computation.

Consider the following illustration.

**Illustration 10.1** The survival times (months) of 10 individuals in a follow-up study are given as 3,4,4,6,8,8,8,10,10,12. We note that for each individual the outcome has to be recorded along with survival time. In this case every outcome is '1' indicating death. Putting these things into MS-Excel we get the data and the corresponding calculations as shown in Table 10.1.

TABLE 10.1: Computation of survival function of individuals with complete follow-up

| S.No | 1 | 2 | 3 | 4 | 5 | 6 | 7 | 8 | 9 | 10 |
|---|---|---|---|---|---|---|---|---|---|---|
| Outcome | 1 | 1 | 1 | 1 | 1 | 1 | 1 | 1 | 1 | 1 |
| Survival time (t) | 3 | 4 | 4 | 6 | 8 | 8 | 8 | 10 | 10 | 12 |
| Rank (i) | 1 | 3 | 3 | 4 | 7 | 7 | 7 | 9 | 9 | 10 |
| Surviving (n-i) | 9 | 7 | 7 | 6 | 3 | 3 | 3 | 1 | 1 | 0 |
| S (t) | 0.9 | 0.7 | 0.7 | 0.6 | 0.3 | 0.3 | 0.3 | 0.1 | 0.1 | 0 |

The row titled 'outcome' shows '1' to all 10 cases. This type of data preparation is necessary while using standard software for survival analysis. The first case gets rank = 1 but the second and third cases have survival time of 4 months and hence the rank for these two cases will be 3. Applying (Equation 10.1) on a simple MS-Excel sheet produces the survival function S(t) as shown in the last row of Table 10.1.

We observe that the survival probabilities decrease as survival time increases. In other words, longer survival times are associated with lower probability.

This analysis can be done conveniently with the help of MedCalc using the following sequence of operations. Statistics → Survival Analysis → Kaplan–Meier survival analysis. Select the 'survival time' as the variable to measure the time and the variable 'Outcome' as the End Point. Using the default setting we get

a) Mean survival months $= 7.3$ , 95% CI : $[5.45, 9.15]$.

b) Median survival months $= 8.0$, 95% CI : $[4.00, 10.00]$.

Unless required otherwise, the median survival time is used as the estimate.

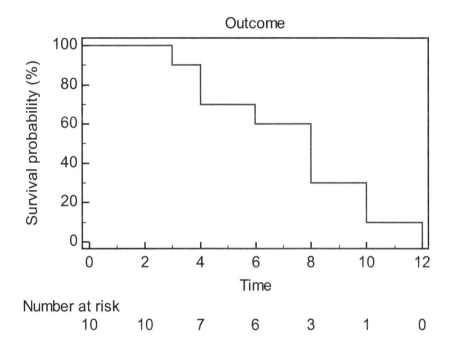

FIGURE 10.1: Survival function plot of 10 individuals.

A quick interpretation from the data is that most of the individuals under study will have a survival time between 4 and 10 months. The survival function is shown in Figure 10.1. The number of individuals at risk was initially 10 and later rapidly decreased to zero, since all were dead. The survival plot is used as a 'visual' for quick understanding of the estimated survival times.

In the following section we discuss a method of estimating S(t) for censored data.

## 10.4   The Kaplan–Meier Method for Censored Data

When the survival times of individuals are not completely followed up during the study period, the data becomes incomplete. The Kaplan–Meier method of estimating the median survival time is used for this type of data. It is based on conditional probabilities and the following formula is used.

Let $S(i)$ = Proportion of persons surviving at $i^{th}$ year. This is estimated from sample data using the relationship

$$\widehat{S}(i) = \widehat{S}(i-1)p_i \tag{10.2}$$

where $p_i$ = Proportion of individuals surviving in the $i^{th}$ year, after they have survived $(i-1)$ years and indicates that the value is a sample estimate.

This is a recurrence formula and the estimate is known as the *Product-Limit* (PL) estimate because it follows from Equation 10.2 that

$$\widehat{S}(i) = p_1 * p_2 * \ldots * p_k \tag{10.3}$$

The computations can again be performed easily with MedCalc as shown in Illustration 10.2.

**Illustration 10.2** The data given in Table 10.2 show the survival times (in weeks) of 30 patients along with the outcome. Those followed until death receive outcome code '1' and the others receive '0'. We wish to estimate the mean survival time.

TABLE 10.2: Survival times (weeks) of patients

| Patient ID | Survival Weeks | Outcome | Patient ID | Survival Weeks | Outcome |
|---|---|---|---|---|---|
| 1 | 9 | 0 | 16 | 12 | 0 |
| 2 | 14 | 0 | 17 | 14 | 1 |
| 3 | 10 | 0 | 18 | 10 | 0 |
| 4 | 12 | 1 | 19 | 12 | 0 |
| 5 | 8 | 0 | 20 | 2 | 1 |
| 6 | 10 | 0 | 21 | 14 | 1 |
| 7 | 8 | 0 | 22 | 3 | 1 |
| 8 | 8 | 0 | 23 | 15 | 0 |
| 9 | 11 | 0 | 24 | 2 | 1 |
| 10 | 6 | 0 | 25 | 19 | 0 |
| 11 | 8 | 0 | 26 | 10 | 0 |
| 12 | 4 | 1 | 27 | 3 | 1 |
| 13 | 5 | 0 | 28 | 7 | 0 |
| 14 | 4 | 1 | 29 | 17 | 1 |
| 15 | 38 | 0 | 30 | 53 | 0 |

In the above data, patients marked with outcome code = 1 were not followed up till the event (death) occurred. For instance patient number 5 was known to survive only for 8 weeks while the survival of patient number 15 was known only up to 38 weeks. Proceeding with MedCalc we get the following results.

a) Mean survival months = 28.0 , 95% CI : [16.0, 39.9].

b) Median survival months = 17.0, 95% CI : [14.0, 17.0].

The survival plot is produced with 95% confidence intervals and the censored cases were displayed with a small vertical dash. The MedCalc options are shown in Figure 10.2.

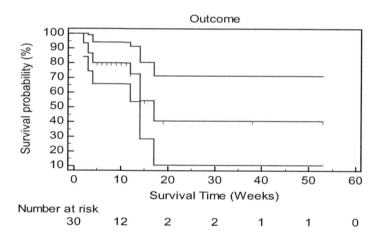

FIGURE 10.2: MedCalc options for Kaplan–Meier survival time estimation.

The plot of the survival function shows the estimated proportion of cases serving at different time points. As an option we can plot the confidence interval for the plot. The plot is shown in Figure 10.3.

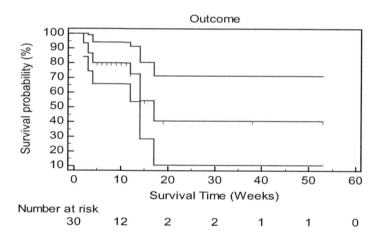

FIGURE 10.3: Survival plot with 95% CI.

We see that the survival times are estimated with wider and wider confidence intervals, which means the estimates are not precise when longer survival times are in question.

**Remark-1:**

The Kaplan–Meier method of estimation can also be used to compare the median survival times between two groups, say placebo and intervention groups. Further the significance of the difference between the median survival times can be tested using a log-rank test. All these options are available in MedCalc as well as SPSS. The chief drawback of this method is that we can estimate the mean or median survival time but only one covariate (factor), eg., treatment group can be handled in the estimation process.

In the following section we deal with a multivariate tool known as *Cox Regression* or *Cox Proportional Hazards* model to estimate the survival times.

---

## 10.5   Cox Regression Model

The very concept of *survival* is linked to *hazard* or *risk* of occurrence of an event. Every individual will have a baseline hazard or risk. From time to time the risk changes and the rate of change over the baseline risk is the key output in survival analysis. This change is often influenced by several factors and an estimate of the chance of survival can be made through regression analysis.

Cox regression is a mathematical model to estimate the *relative risk* or *risk change* taking into account a) baseline risk and b) the influence of covariates. It is based on the hazard rate $h(t)$ which is defined as probability that an individual of age 't' dies in the interval $(t, t+\delta t)$ where $\delta t$ is a small incremental change in time. The function $h(t)$ is called *instantaneous hazard or force of mortality*. The hazard rate could be a constant or it could be increasing or decreasing with time.

In Cox regression we wish to estimate the conditional risk of death for an individual at time t, given the values of the covariates $(x_1, x_2, \ldots, x_k)$. This is modeled as $h(t \mid x_1, x_2, \ldots, x_p) = h_0(t)e^{(\beta_1 x_1 + \beta_2 x_2 + \ldots + \beta_k x_k)}$ where $h_0(t)$ is called the *baseline hazard*.

More specifically the Cox model for the $i^{th}$ individual is stated as

$$h_i(t \mid x_{1i}, x_{2i}, \ldots, x_{ki}) = h_0(t)e^{(\beta_1 x_{1i} + \beta_2 x_{2i} + \ldots + \beta_k x_{ki})} \qquad (10.4)$$

where for the $i^{th}$ individual, $(x_{1i}, x_{2i}, \ldots, x_{ki})$ denote the values of covariates, $\beta_1, \beta_2, \ldots, \beta_p$ denote the regression coefficients and $h_i(t \mid x_{1i}, x_{2i}, \ldots, x_{pi})$ denote the hazard at time t. With a simple transformation, the model in Equation 10.4 can be written as

$$\ln\left[\frac{h_i(t \mid (x_{1i}, x_{2i}, \ldots, x_{ki})}{h_0(t)}\right] = \beta_1 x_{1i} + \beta_2 x_{2i} + \ldots + \beta_k x_{ki} \qquad (10.5)$$

If we denote the left-hand side of Equation 10.4 by Y then Cox model becomes a multiple linear regression model where Y denotes the *log of hazard ratio*.

If we denote $S_{(t)}$ and $S_0(t)$ as the survival probability at t and at baseline respectively then it can be shown that Equation 10.5 reduces to

$$S(t) = S_0(t)^k \text{ where } k = e^{\sum_{j=1}^{k} \beta_j x_j} \tag{10.6}$$

The coefficients $(\beta_1, \beta_2, \ldots, \beta_k)$ are estimated by maximum likelihood method basing on the conditional log likelihood. This is a data-driven exercise and k-equations have to be solved to estimate the coefficients.

Standard software like R, SPSS or SAS can handle the computations. Eric Vittinghoff *et.al* (2004) contains interesting illustrations for survival analysis. MedCalc also has a module to handle Cox regression. The output of most software contains the estimates of $\beta_1, \beta_2, \ldots, \beta_k$, their standard errors and 95% confidence intervals. SPSS however has an option to save, for each case, the predicted values of $\widehat{S}(t)$, the hazard function h(t) and the score $(X*\beta)$ obtained from the right hand side of Equation 10.6. As a visual aid, the survival or hazard function can be plotted. They can also be compared between groups (e.g., gender or treatment arms).

In the following section we illustrate the method with MedCalc.

**Illustration 10.3** Reconsider the ICU scores data used in Illustration 7.1. Table 10.3 contains a portion of data with 15 records. The variables are age, gender, APACHE score, DurHospStay and the outcome. The outcome variable is coded as '1' when the patient is dead and 0 when discharged (censored). The analysis is however performed on the first 64 records of the original data.

TABLE 10.3: Hospital stay data

| S.NO | Age | Gender | APACHE | DurHospStay | Outcome |
|------|-----|--------|--------|-------------|---------|
| 1 | 20 | 0 | 12 | 12 | 0 |
| 2 | 52 | 1 | 21 | 12 | 1 |
| 3 | 25 | 1 | 20 | 9 | 0 |
| 4 | 53 | 1 | 20 | 10 | 0 |
| 5 | 15 | 1 | 15 | 14 | 0 |
| 6 | 40 | 1 | 21 | 10 | 0 |
| 7 | 70 | 1 | 16 | 7 | 0 |
| 8 | 50 | 1 | 22 | 12 | 1 |
| 9 | 27 | 1 | 9 | 9 | 0 |
| 10 | 30 | 0 | 13 | 22 | 0 |
| 11 | 47 | 1 | 23 | 12 | 1 |
| 12 | 23 | 0 | 15 | 8 | 0 |
| 13 | 19 | 0 | 8 | 10 | 0 |
| 14 | 40 | 0 | 11 | 9 | 0 |
| 15 | 30 | 0 | 21 | 4 | 1 |

The researcher wants to study the effect of age, gender and APACHE score on the duration of hospital stay (in ICU) until either discharged or dead.

**Analysis:**
The MedCalc options will be as follows.

1. Create the data file in MedCalc or create the same in SPSS or MS-Excel and read in MedCalc.

2. Select Statistics → Survival Analysis → Cox proportional hazards regression.

3. Select Survival time → DurHospStay.

4. Select Endpoint → Outcome. The status is by default taken as '1' indicating the event of interest (death).

5. Select one predictor variable, say APACHE, and leave the options at their defaults.

6. Press OK.

The output of Cox regression shows the following information.

a) Cox proportional-hazards regression
   Survival time: DurHospStay

Endpoint: Outcome
Method: Enter

Since there is only one predictor variable, the regression method is taken as 'Enter' (we may choose *forward conditional* if more than one predictor variable is proposed).

b) Case summary

Number of events (Outcome = 1): 20 (31.25%)
Number censored (Outcome = 0): 44 (68.75%)
Total number of cases: 64 (100.00%)

About 69% of cases were censored and 31% are the events of interest.

c) Overall model fit

-2 Log Likelihood =150.874, Chi-squared = 28.722,DF= 1, p < 0.0001

A measure of goodness of fit is the index -2logLikelihood and the *fit by chance* is rejected since the p-value is much smaller than the level of significance (p < 0.01) basing on the Chi-square test. Hence the model is a good fit.

d) Coefficients and standard errors

| Covariate | b | SE | Wald | p | Exp(b) | 95% CI of Exp(b) |
|---|---|---|---|---|---|---|
| APACHE | 0.1644 | 0.0300 | 30.0867 | <0.0001 | 1.1787 | 1.1118 to 1.2497 |

The predictor APACHE has an estimated coefficient b = 0.1644 which is significantly different from '0' (null hypothesis). The value of Exp(b) indicates the *relative risk* which is similar to *odds ratio*. In this case it means when the APACHE score increases by '1' the risk of death increases by about 17.8%. The confidence interval (CI) indicates that the additional risk over the baseline will be anywhere from 11.18% to 24.97% when the APACHE score increases by one unit.

e) Baseline cumulative hazard function

The summary output of Cox regression is reflected in terms of the predicted survival probability and the cumulative hazard rate, evaluated at the mean values of the covariates (here only one covariate, APACHE). The pattern of survival probability (%) is also an essential output for understanding the steepness in the fall of survival time as the time to event increases.

TABLE 10.4: Survival and cumulative hazard table

| Time | Baseline Cumulative Hazard | At Mean of Covariates | |
| --- | --- | --- | --- |
| | | Cumulative Hazard | Survival |
| 2 | 0.000 | 0.008 | 0.992 |
| 3 | 0.003 | 0.047 | 0.954 |
| 4 | 0.006 | 0.096 | 0.909 |
| 5 | 0.007 | 0.113 | 0.893 |
| 6 | 0.008 | 0.131 | 0.877 |
| 9 | 0.009 | 0.154 | 0.857 |
| 12 | 0.016 | 0.268 | 0.765 |
| 13 | 0.020 | 0.344 | 0.709 |
| 14 | 0.025 | 0.432 | 0.649 |

Table 10.4 shows the survival pattern and Figure 10.4 gives the associated plot.

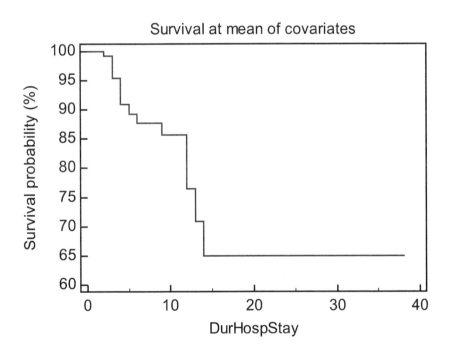

FIGURE 10.4: Survival plot for duration of hospital stay.

From the Cox model it is observed that the APACHE score has a significant impact on the duration of hospital stay with an odds ratio (relative risk) of 1.1788. Suppose the researcher suspects that the risk could be partially influenced by gender or age, then we have to *adjust* the regression coefficients and derive the new *adjusted relative risk*. This is done by simply including age and gender, into the regression as covariates.

Running the tool with these options we get the following values adjusted for age and gender.

a) The reference category is Gender = 0 (Female). Both age and gender had no significant influence on the survival.

b) The adjusted relative risk for APACHE is 1.1847 (CI: 1.1106, 1.2637) (without adjustment, this value was 1.1788).

c) The gender-wise survival plots are displayed in Figure 10.5.

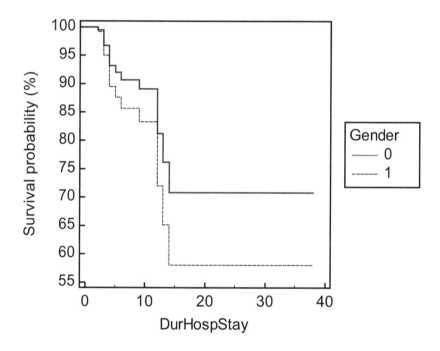

FIGURE 10.5: Survival plot for duration of hospital stay adjusted for age and gender.

It can be seen from Figure 10.5 that when compared to females (code = 0) males have lower survival rate.

**Remark 1**

a) Sometimes the covariates can be time dependent such as waiting time for organ transplantation. Long waiting times before obtaining the organ leads to time-dependent changes in vital parameters.

b) Similarly, some predictors will be ordinal such as histology ratings. Since they are not nominal, the reference values shall be carefully fixed before comparison.

c) In some cases the study cohort itself may have different baseline hazard as in the case of cohorts which are stratified basing on ethnic groups, diet habits etc. Cox regression with a stratified cohort, will addresses this problem.

Survival analysis itself is a specialized subject and needs careful understanding and implementation of computational tools. Powerful algorithms and procedures are available in R, SAS, STATA etc. and the user has to choose the right tool before proceeding with applications.

We end this chapter with the note that Kaplan–Meier estimates and Cox proportional hazard regression are vital tools in survival analysis.

---

## Summary

Survival analysis is a statistical method of estimating the time-to-happening of an event. When the event of interest is dichotomous, we wish to estimate the hazard rate (instantaneous risk of event at a given time point) and the pattern of survival chance in terms of duration. It is used to predict the survival time of patients after a treatment or duration of disease-free survival etc. The Kaplan–Meier method is one popular tool to estimate the mean/median survival time and it is also used to compare the survival pattern between treatment groups. When more than one factor is likely to influence the survival, we can use the Cox regression (proportional hazard) model and estimate the resulting relative risk (of event). Both Kaplan–Meier and Cox regression are computer-intensive methods and software helps in quick and reliable results. Survival analysis is applicable even in the non-clinical context such as insurance, aircraft maintenance etc.

## Do it yourself (Exercises)

**10.1** The following data refers to the survival times (months) of 20 patients who were on dialysis.

| Patient | Survival Months |
|---------|-----------------|
| 1 | 5 |
| 2 | 6 |
| 3 | 6 |
| 4 | 10 |
| 5 | 6 |
| 6 | 12 |
| 7 | 11 |
| 8 | 16 |
| 9 | 21 |
| 10 | 9 |

| Patient | Survival Months |
|---------|-----------------|
| 11 | 7 |
| 12 | 11 |
| 13 | 13 |
| 14 | 16 |
| 15 | 21 |
| 16 | 20 |
| 17 | 20 |
| 18 | 18 |
| 19 | 8 |
| 20 | 10 |

Find the median survival months using the Kaplan–Meier method and plot the survival function.

**10.2** The following data was obtained in a study on clinical management of sarcoma among 22 osteosarcoma patients. The following variables are selected for studying overall survival pattern among the patients.

| Variable | Description |
|----------|-------------|
| X1 | Age in Years |
| X2 | Sex (Male = 1; Female = 2) |
| X3 | Duration of Symptoms (Months) |
| X4 | Tumor Size ($< 5$cm = 1; 5-10cm =2; $>10$cm = 3) |
| X5 | State at Diagnosis (0-non metastatic; 1- metastatic) |
| X6 | Hemoglobin |
| X7 | Albumin |
| X8 | Histology |
| X9 | Outcome (0 = Alive; 1 = Dead) |
| X10 | Overall Survival Days |
| X11 | Overall Survival Months |

| S.No | X1 | X2 | X3 | X4 | X5 | X6 | X7 | X8 | X9 | X10 | X11 |
|------|----|----|----|----|----|------|-----|----|----|------|------|
| 1 | 15 | 2 | 4 | 2 | 0 | 9.8 | 3.2 | 2 | 0 | 1815 | 60.5 |
| 2 | 17 | 1 | 3 | 2 | 0 | 10.1 | 3.6 | 1 | 0 | 186 | 6.2 |
| 3 | 9 | 2 | 2 | 3 | 0 | 11.6 | 4.2 | 1 | 0 | 1660 | 55.3 |
| 4 | 16 | 2 | 2 | 2 | 0 | 11.7 | 3.0 | 2 | 0 | 548 | 18.3 |
| 5 | 18 | 1 | 4 | 3 | 0 | 15.8 | 4.2 | 1 | 1 | 878 | 29.3 |
| 6 | 18 | 1 | 3 | 3 | 0 | 8.9 | 3.4 | 1 | 0 | 287 | 9.6 |
| 7 | 17 | 1 | 3 | 2 | 0 | 15.1 | 3.1 | 1 | 0 | 1397 | 46.6 |
| 8 | 14 | 1 | 2 | 2 | 0 | 11.4 | 4.1 | 2 | 0 | 474 | 15.8 |
| 9 | 15 | 2 | 4 | 2 | 0 | 9.8 | 3.7 | 1 | 0 | 2013 | 67.1 |
| 10 | 9 | 2 | 1 | 2 | 0 | 12.2 | 3.2 | 1 | 0 | 1010 | 33.7 |
| 11 | 18 | 1 | 3 | 2 | 0 | 16.0 | 3.7 | 1 | 0 | 636 | 21.2 |
| 12 | 11 | 2 | 3 | 2 | 0 | 10.7 | 3.9 | 1 | 0 | 332 | 11.1 |
| 13 | 18 | 1 | 2 | 3 | 0 | 13.5 | 3.2 | 3 | 0 | 972 | 32.4 |
| 14 | 16 | 2 | 5 | 3 | 0 | 9.6 | 3.7 | 2 | 0 | 62 | 2.1 |
| 15 | 15 | 1 | 3 | 3 | 0 | 12.8 | 3.3 | 1 | 1 | 131 | 4.4 |
| 16 | 13 | 2 | 2 | 2 | 0 | 12.2 | 3.9 | 2 | 1 | 601 | 20.0 |
| 17 | 18 | 1 | 3 | 2 | 0 | 13.7 | 4.4 | 1 | 0 | 606 | 20.2 |
| 18 | 13 | 1 | 2 | 2 | 0 | 12.8 | 3.7 | 1 | 0 | 127 | 4.2 |
| 19 | 12 | 2 | 3 | 2 | 0 | 12.0 | 4.5 | 1 | 0 | 267 | 8.9 |
| 20 | 5 | 2 | 1 | 3 | 0 | 7.1 | 3.5 | 2 | 1 | 752 | 25.1 |
| 21 | 18 | 2 | 6 | 2 | 0 | 12.2 | 4.2 | 1 | 0 | 338 | 11.3 |
| 22 | 16 | 2 | 2 | 3 | 0 | 10.7 | 3.3 | 1 | 0 | 596 | 19.9 |

    a) Obtain Kaplan–Meier survival function using overall survival months.

    b) Compare the median survival time between males and females and comment on the significance of the difference.

**10.3** Use the data given in Exercise 10.2 and obtain the survival plots for different tumor sizes using overall survival months.

**10.4** Use the data given in Exercise 10.2 and estimate the overall survival days using Cox regression with relevant predictors. (Hint: use the stepwise method.)

**10.5** Use MedCalc to obtain the cumulative hazard function after adjusting for age and gender for the data given in Exercise 10.2 by taking end point = outcome and survival time = overall survival months.

**10.6** The following data refers to the survival time (in days) of 50 leukemia patients and the variable information is given in the table below.

| Variable | Description |
|----------|-------------|
| X1 | Sex (Male = 1; Female = 0) |
| X2 | Risk (1 = High; 2 = Standard |
| X3 | Outcome (0 = Alive; 1 = Dead) |
| X4 | Survival (in days) |

| S.No | X1 | X2 | X3 | X4 | S.No | X1 | X2 | X3 | X4 |
|------|----|----|----|------|------|----|----|----|------|
| 1 | 0 | 1 | 0 | 528 | 26 | 0 | 1 | 1 | 10 |
| 2 | 0 | 1 | 0 | 22 | 27 | 1 | 2 | 1 | 15 |
| 3 | 0 | 1 | 0 | 109 | 28 | 0 | 1 | 0 | 363 |
| 4 | 1 | 2 | 0 | 1125 | 29 | 1 | 1 | 0 | 29 |
| 5 | 0 | 1 | 1 | 23 | 30 | 0 | 2 | 0 | 190 |
| 6 | 0 | 1 | 0 | 345 | 31 | 1 | 1 | 0 | 305 |
| 7 | 0 | 2 | 0 | 970 | 32 | 1 | 1 | 0 | 658 |
| 8 | 0 | 1 | 1 | 6 | 33 | 0 | 2 | 1 | 22 |
| 9 | 0 | 1 | 0 | 240 | 34 | 0 | 1 | 0 | 117 |
| 10 | 0 | 2 | 1 | 23 | 35 | 0 | 1 | 0 | 1005 |
| 11 | 1 | 1 | 0 | 233 | 36 | 1 | 1 | 0 | 517 |
| 12 | 0 | 1 | 0 | 472 | 37 | 1 | 2 | 0 | 515 |
| 13 | 1 | 1 | 0 | 601 | 38 | 0 | 1 | 0 | 118 |
| 14 | 1 | 1 | 0 | 920 | 39 | 0 | 1 | 1 | 2 |
| 15 | 1 | 2 | 0 | 916 | 40 | 0 | 1 | 0 | 213 |
| 16 | 1 | 1 | 0 | 713 | 41 | 0 | 2 | 0 | 319 |
| 17 | 0 | 1 | 0 | 742 | 42 | 1 | 1 | 1 | 4 |
| 18 | 1 | 1 | 0 | 888 | 43 | 1 | 1 | 0 | 318 |
| 19 | 0 | 1 | 1 | 28 | 44 | 0 | 1 | 0 | 152 |
| 20 | 0 | 2 | 0 | 849 | 45 | 0 | 1 | 1 | 16 |
| 21 | 0 | 2 | 0 | 103 | 46 | 0 | 2 | 0 | 48 |
| 22 | 0 | 1 | 0 | 94 | 47 | 0 | 2 | 0 | 160 |
| 23 | 1 | 1 | 1 | 12 | 48 | 1 | 2 | 0 | 487 |
| 24 | 0 | 2 | 0 | 53 | 49 | 0 | 2 | 0 | 104 |
| 25 | 0 | 2 | 0 | 351 | 50 | 0 | 1 | 0 | 260 |

a) Perform Kaplan–Meier analysis using survival days and draw conclusions.

b) Perform the same using R-code by selecting the appropriate library functions.

**10.7** Use the data given in Exercise 10.5 and obtain the risk estimates using Cox proportional hazards model taking gender and level of risk as predictors and interpret the findings.

# Suggested Reading

1. Lee, E.T. 2013. *Statistical Methods for Survival Data Analysis.* $4^{\text{th}}$ ed. John Wiley & Sons, Inc.

2. Eric Vittinghoff, Stephen C. Shiboski, David V. Glidden, Charles E. McCulloch. 2004. *Regression Methods in Boistatisitcs-Linear, Logistic, Survival and Repeated Measures Models*: Springer.

3. Kaplan, E.L and Paul Meier. 1958. Nonparametric Estimation from Incomplete Observations. *Journal of the American Statistical Association*, 53 (282), 457-481.

# Chapter 11

## Poisson Regression Analysis

11.1    Introduction ..................................................... 205
11.2    General Form of Poisson Regression ........................... 206
11.3    Selection of Variables and Subset Regression .................. 210
11.4    Poisson Regression Using SPSS ............................... 212
11.5    Applications of the Poisson Regression model ................. 216
       Summary ..................................................... 217
       Do it yourself (Exercises) ..................................... 217
       Suggested Reading .............................................. 218

> Not everything that can be counted counts, and not every-
> thing that counts can be counted.
>
> Albert Einstein (1879 – 1955)

## 11.1   Introduction

In multiple linear regression the response/outcome variable (Y) is on a measurement scale which means Y is continuous. When the outcome is nominal, we use binary or multinomial logistic regression or a linear discriminant function. However there are instances where the response variable accounts for the data which is the count/frequency of an event observed during an experiment or survey. In a biological experiment, the occurrence of a particular kind of bacteria in a colony, number of pediatric patients observed with Acute Myeloid Leukemia (AML), number of machines with a minute dysfunctionality are some examples. Values of such data will be 0,1,2,... known as *count data*. Statistically the word 'count' refers to "the number of times an event occurs and represented by a non-negative integer valued random variable."

The occurrence of events can be viewed in a probabilisticframe work with a distribution to explain the pattern. The notable distributions that possess the features of count data are binomial and poisson. Of these two, Poisson distribution emphasizes particular type of count data in rare occurrences. Theoretically, Poisson (1837) viewed this distribution as a limiting case of binomial distribution and Greenwood and Yule (1920) gave the standard generalisation of Poisson and called it negative binomial distribution. The main focus of working with count data is to address the issues such as heterogeneity and overdispersion. In general, applications for count data analysis are observed in diverse fields such as economics, biostatistics, demography, actuarial science etc.

In estimating the probability of count data, the support of several co-variates (explanatory variables) is essential. This leads to forming a linear combination with a set of covariates that help in estimating the probability of the event (outcome) of interest and such a model is called *count data regression*. Over the years, umpteen number of articles were published on count data regression models that led to the development of an emerging area namely, *Generalized Linear Models*. Among many count data regression models, the Poisson model is treated as a special case and a specific form of non-linear regression that accounts for the characteristic of discreteness of a count variable. Cameron and Trivedi (1998) detailed the usefulness of count data regression in studying the occurrence rate per unit of time conditional on some covariates. Apart from its application to longitudinal or time series data, cross-sectional data is also considered to apply to Poisson regression.

In the following section the mathematical model of Poisson regression is explained, followed by an illustration.

---

## 11.2     General Form of Poisson Regression

Let Y be a random variable observed with Poisson events. The probability of occurrence of such events can be obtained as

$$P(Y = y) = \begin{cases} \frac{e^{-\mu}\mu^y}{y!} & ; y = 0, 1, 2, \ldots \\ 0 & ; otherwise \end{cases} \tag{11.1}$$

where $\mu$ is the mean number of occurrences of the event of interest. For this distribution, $E(Y) = V(Y) = \mu$, which is known as the *equidispersion property* , means that average and variance are equal for Y.

Let $X_1, X_2, \ldots, X_k$ be k predictors and $\beta_0, \beta_1, \ldots, \beta_k$ be coefficients such that the outcome Y is a function of $X_1, X_2, \ldots, X_k$. We assume that Y can be described by a Poisson distribution with parameter $\mu$, given in Equation 11.1.

In Poisson regression, estimate the average of Y given the values of the predictors. For the $i^{th}$ individual, $E(Y_i) = \mu_i$ denotes the expected number of 'events' conditional on the predictor values $X_{1i}, X_{2i}, \ldots, X_{ki}$.

As such, the model for Poisson mean is given by

$$\mu(X_i, \beta_i) = \exp\{ \beta_0 + \sum_{i=1}^{k} \beta_j X_j \} \forall\, i = 1, 2, \ldots, n$$

The parameters $\beta_0, \beta_1, \ldots, \beta_k$ are estimated with the maximum likelihood procedure. The ordinary least square method can not be applied since Y is discrete. The goodness of fit of Poisson regression models can be obtained using the deviance statistic approximated to Chi-square distribution with (n-p-1) degrees of freedom.

In this chapter, the application of Poisson regression is demonstrated using real as well as simulated datasets with the R platform and SPSS.

Consider the following illustration.

**Illustration 11.1** Consider a study on subjects who were diagnosed with oral cancer to estimate the number of infected lymph nodes in terms of several prognostic factors viz. Tumor Thickness (TT), Resistivity Index (RI), Pulsatality Index (PI), Systolic Pressure (PS), Diastolic Pressure (ED) and Number of Lymph Nodes infected (LN). A portion of data with 15 samples is shown in Table 11.1 out of 45 samples. The data analysis however is carried out on the complete dataset.

TABLE 11.1: Lymp node data

| S.No | TT | RI | PI | PS | ED | LN |
|------|------|------|------|-------|-------|----|
| 1 | 1.68 | 0.52 | 0.89 | 27.30 | 13.10 | 2 |
| 2 | 1.38 | 0.51 | 0.62 | 19.73 | 9.67 | 3 |
| 3 | 1.57 | 0.32 | 0.63 | 32.14 | 21.85 | 3 |
| 4 | 1.71 | 0.48 | 0.81 | 20.61 | 10.72 | 2 |
| 5 | 0.75 | 0.54 | 0.78 | 21.58 | 9.92 | 0 |
| 6 | 0.80 | 0.58 | 0.83 | 18.15 | 7.63 | 0 |
| 7 | 1.50 | 0.47 | 0.92 | 27.65 | 14.66 | 2 |
| 8 | 2.62 | 0.33 | 0.94 | 28.42 | 19.04 | 5 |
| 9 | 2.06 | 0.55 | 0.96 | 20.06 | 9.02 | 5 |
| 10 | 1.05 | 0.58 | 1.30 | 16.82 | 7.06 | 1 |
| 11 | 0.59 | 0.65 | 1.71 | 26.80 | 9.38 | 0 |
| 12 | 1.91 | 0.41 | 0.87 | 23.91 | 14.10 | 3 |
| 13 | 1.61 | 0.51 | 0.82 | 23.80 | 11.66 | 2 |
| 14 | 2.83 | 0.42 | 0.79 | 23.07 | 13.39 | 6 |
| 15 | 1.43 | 0.56 | 0.92 | 20.34 | 8.94 | 2 |

We wish to develop a model to predict LN count in terms of covariates using the Poisson regression model.

**Analysis:**

The analysis is carried out using R. The prime objective is to estimate the number of lymph nodes given the information on the predictors. The following are the sequence of R codes used for building and understanding the model behaviour.

1. *pr=read.csv(file.choose(), header = TRUE)*
   The data will be stored as a new variable 'pr'. The read.csv( ) command will help us to import the data into the R environment. Other formats such as '.txt', '.dat', '.xls' are also functional in the place of '.csv'.

2. The *attach(pr)* command is used to access the objects stored in 'pr' directly for use in other formats.

3. The *head(pr)* command will display the first six records (by default) of the dataset as shown below

| S.No | TT | RI | PI | PS | ED | LN |
|------|------|------|------|-------|-------|----|
| 1 | 1.68 | 0.52 | 0.89 | 27.30 | 13.10 | 2 |
| 2 | 1.38 | 0.51 | 0.62 | 19.73 | 9.67 | 3 |
| 3 | 1.57 | 0.32 | 0.63 | 32.14 | 21.85 | 3 |
| 4 | 1.71 | 0.48 | 0.81 | 20.61 | 10.72 | 2 |
| 5 | 0.75 | 0.54 | 0.78 | 21.58 | 9.92 | 0 |
| 6 | 0.80 | 0.58 | 0.83 | 18.15 | 7.63 | 0 |

4. To run the Poisson regression, the glm( ) function is used. All the output produced by glm( ) can be stored into a temporary variable (say 'model') for further analysis. Here LN is the response variable and family to be chosen as 'poisson' with 'log' as link function. The command is
   *model=glm(LN ~ TT + RI + PI + PS + ED, family=poisson(link=log), data=pr)*

5. If we type *summary(model)* and press 'ctrl+R', we get output as follows

   a) Deviance Residuals:

| Min | 1Q | Median | 3Q | Max |
|--------|--------|--------|-------|-------|
| -1.946 | -0.428 | -0.081 | 0.426 | 1.457 |

   b) Coefficients:

|              | Estimate | Std.Error | Z value | Pr(>\|z\|) |
| ------------ | -------- | --------- | ------- | ---------- |
| (Intercept)  | 0.020    | 2.537     | 0.008   | 0.994      |
| TT           | 0.256    | 0.187     | 1.366   | 0.172      |
| RI           | 2.409    | 5.009     | 0.481   | 0.631      |
| PI           | -0.295   | 0.476     | -0.620  | 0.535      |
| PS           | -0.117   | 0.114     | -1.024  | 0.306      |
| ED           | 0.181    | 0.195     | 0.929   | 0.353      |

(Dispersion parameter for Poisson family taken to be 1).
Null deviance: 47.382 on 44 degrees of freedom.
Residual deviance: 31.834 on 39 degrees of freedom.
Number of Fisher scoring iterations:5; AIC:155.39.

6. **Model:** The Poisson Regression (PR) model is then given as:

$$\ln(\mu) = 0.0197 + 0.2558 * TT + 2.40875 * RI$$
$$-0.2954 * PI - 0.1167 * PS + 0.1810 * ED$$

Substituting for TT, RI, PI, PS and ED we get a value $Y_i$ which is denoted $\ln(\mu_i)$ so that $\hat{\mu}_i = e^{Y_i}$ is the predicated value (mean). This values should be rounded to make it an integer.

For each predictor, the coefficient (the value before * in the above model) represents the marginal contribution of that factor to $\ln(\mu)$. If we observe the estimates column, the parameter 'RI' has the maximum coefficient value and next to it PI and TT. The higher the coefficient value of a predictor, the larger contribution in predicting the outcome.

7. **Relative Risk (RR):** The relative risk of a predictor is given by exp(B), where B denotes the regression coefficient of that predictor. The RR for each predictor is obtained and shown below.

| Predictor | TT    | RI     | PI     | PS     | ED    |
| --------- | ----- | ------ | ------ | ------ | ----- |
| B         | 0.256 | 2.409  | -0.295 | -0.117 | 0.181 |
| Exp(B)    | 1.291 | 11.120 | 0.744  | 0.890  | 1.198 |

For instance, in the case of RI, the relative risk is 11.1199, which means that one unit increase in the value of RI leads to an 11 fold increase in LN.

8. **Goodness of Fit:** The model fit can be assessed using either $R^2$ or the 'Deviance Statistic.' Here, we choose deviance for observing the model fit. The code *"1-pchisq(model\$deviance,model\$df.residual)"* gives p-value $= 0.7853$. Thus $p > 0.05$ model can be considered as a good fit, because the null hypothesis is 'deviance=0' and it is accepted. Another way is

to take the ratio of residual deviance and degrees of freedom and if it is less than or equal to '1', it indicates adequacy of the model.

Here, the residual deviance and degrees of freedom are 31.834 and 39 respectively, the ratio of these two turns out to be 0.8162 < 1. Hence the model is adequate.

9. **Predicted Values:** On using the R code given below, we obtain the fitted values (number of lymph nodes) for each case using the model. If we observe the first record, by substituting the data values in the above PR model, the expected number of lymph nodes predicated by the model is 2.1164901 or 2 nodes. Similarly, we obtain the fitted values for other records and a sample of 15 records show in Table 11.2

**R Code:** *data.frame(pr,pred=model$ fitted)*

TABLE 11.2: Predicted number of Lymph Nodes (LN) using the PR model

| S.No | Observed | Predicted | Residual |
|------|----------|-----------|----------|
| 1 | 2 | 2.1164 | 2 |
| 2 | 2 | 2.0262 | 2 |
| 3 | 1 | 1.8507 | 2 |
| 4 | 3 | 2.7172 | 3 |
| 5 | 5 | 3.8018 | 4 |
| 6 | 0 | 1.7473 | 2 |
| 7 | 2 | 1.8647 | 2 |
| 8 | 3 | 3.3493 | 3 |
| 9 | 5 | 2.4075 | 2 |
| 10 | 0 | 1.8934 | 2 |
| 11 | 0 | 0.8189 | 1 |
| 12 | 2 | 2.4800 | 2 |
| 13 | 6 | 3.4980 | 3 |
| 14 | 2 | 1.9922 | 2 |
| 15 | 3 | 2.3748 | 2 |

In the following section, we deal with identifying a few important predictors that contribute a better way to predict the outcome. Using the data given in Illustration 11.1, the utility of variable selection is presented.

## 11.3    Selection of Variables and Subset Regression

It is always a good practice to identify a subset of predictors that really provide good prediction of the outcome. In general the model fit shows adequacy with increasing number of predictors even though all of them are not required in the model. So, we search for the subsets of predictors which give a satisfactory fit. To do so, we have to make use of certain procedures which will support listing the best subset of predictors in the R environment.

Here is an illustration.

**Illustration 11.2** Reconsider the data given in Illustration 11.1. We wish to select a subset of variables (instead of all) the data for which best predicts the number of lymph nodes by using the dot on predictors.

**Analysis:**

To have a proper screening of the predictors, we use a "library(FWDselect)," which supports the procedure for a forward selection of variables sequentially into the model. In order to install any package, we make use of "install.packages( )" and choose the option 'install dependencies'. This FWDselect package has two options.

**q-method:**

The code given below returns the best subset and the results purely depend on the choice of 'q', which denotes the desired size of the subset. For instance with q = 2, we get the following combinations for subset of size 2

{(TT,RI), (TT,PI), (TT,PS), (TT,ED), (RI,PI), (RI, PS), (RI, ED), (PI, PS), (PI, ED) and (PS,ED)}

If we take q = 3, all possible subsets of size 3 will be evaluated.

We can have the R code to choose the best as shown below.

**R Code:**

```
library(FWDselect)
vs1=selection(x=pr[,6],y=pr[,6],q=2,method="glm",family=poisson(link=log),
criterion="deviance)
vs1
(press ctrl+R)
```

For instance, if we take q=2, the function will return the best possible combination of (size 2) predictors as shown below. Here RI and TT are found to be a significant combination towards the response variable among all the paired combinations of predictors.

```
**************************************************
```
Best subset of size q = 2 : RI TT
Information Criterion Value - deviance : 8.483402

```
**************************************************
```

**Vector method:**

However, another way of screening the variables is to use a vector that lists all possible predictor combinations of different sizes. The code for such execution is given below.

**R Code:**

```
vs2=qselection(x=pr[,-6],y=pr[,6],qvector=c(1:4),method="glm",
family=poisson(link=log),criterion="deviance")
vs2
```
(press ctrl+R)

The argument 'qvector=c(1:4)' will generate all possible combinations of size 1, 2, 3 and 4 of which, the function iteratively selects the best subset for each q.

Table 11.3 depicts the final result of subset selection. The symbol '*' is automatically generated by the function to signal that among the selected subset sizes, the size q=1, i.e., the predictor RI is the one that can significantly contribute in predicting the number lymph nodes. Among the possible subsets we select the one having the smallest deviance.

TABLE 11.3: Screening of variables by vector method

| q | Deviance | Selection |
|---|----------|-----------|
| 1 | 7.921 | RI* |
| 2 | 8.483 | RI, TT |
| 3 | 8.775 | RI, TT, PI |
| 4 | 9.656 | ED, TT, PI, PS |

In the following section, the options that are available in SPSS to perform the PR model are explored.

## 11.4   Poisson Regression Using SPSS

SPSS has a module with user-friendly menus and options for running a Poisson regression, available in the 'Generalized linear models' tab in the Analyze menu. we can also get the predicted values and the residuals can be saved to file. Here is an illustration.

**Illustration 11.3** Reconsider the ICU scores data used in Illustration 11.1. A sample of the first 10 records with variables i) number of days spent on ventilator (days of vent), ii) Shock (1 = Yes, 0 = No), iii) Diagnosis (Dia2: 1 = Sepsis, 2 = Severe Sepsis, 3 = Septic shock) and iv) duration of hospital stay (DurHospStay) is shown in the Table 11.4.

TABLE 11.4: Data for Poisson regression

| S.No | Days of Vent | Shock | Diagnosis | DurHospStay |
|------|------|------|------|------|
| 1 | 5 | 1 | 3 | 12 |
| 2 | 5 | 0 | 2 | 12 |
| 3 | 0 | 0 | 2 | 9 |
| 4 | 3 | 1 | 2 | 10 |
| 5 | 5 | 1 | 2 | 14 |
| 6 | 4 | 1 | 3 | 10 |
| 7 | 2 | 0 | 2 | 7 |
| 8 | 5 | 1 | 2 | 12 |
| 9 | 0 | 1 | 3 | 9 |
| 10 | 18 | 1 | 2 | 22 |
| 11 | 12 | 1 | 3 | 12 |
| 12 | 4 | 0 | 2 | 8 |
| 13 | 0 | 0 | 2 | 10 |
| 14 | 4 | 0 | 2 | 9 |
| 15 | 4 | 1 | 3 | 4 |

The analysis is however done using the first 50 records of the original dataset.

We wish to estimate the number of days on a ventilator in terms of the predictors shock, diagnosis and duration of hospital stay.

**Analysis:**

The analysis is carried out in SPSS with the following steps.

1. Analyze → Generalized Linear Models → Generalized Linear Models.

2. Under the type of Model, choose 'Poisson loglinear'.

3. In the Response tab, send the response variable i.e., days of vent to the 'Dependent Variable' Pane as shown in Figure 11.1.

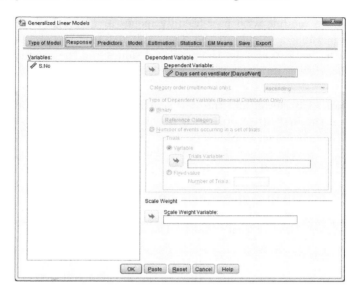

FIGURE 11.1: SPSS option for Poisson regression.

4. Move to the Predictors tab for selecting Diagnosis and Shock as 'Factors' and DurHospStay as 'Covariates'.

5. Go to the Model tab, and send the variables Diagnosis, Shock and DurHospStay to the 'Model' pane.

6. Click on the 'Save' tab and select 'Predicated value of mean of response' and 'Residual' as 'Item to Save'.

7. Press OK to view the results.

The output can be understood and documented as follows.

**Variable significance:**

Table 11.5 below contains the result of testing the significance of each variable, chosen for estimating the days on vent. From this, we can observe that except for 'Diagnosis', the other two are statistically significant at $p < 0.05$. It means that each of them shows a significant variable has a their marginal effect on the response variable.

TABLE 11.5: Tests of model effects

| Source | Wald Chi-Square | df | Sig. |
|---|---|---|---|
| (Intercept) | 17.458 | 1 | 0.000 |
| Diagnosis | 4.493 | 2 | 0.106 |
| Shock | 3.941 | 1 | 0.047 |
| DurHospStay | 99.949 | 1 | 0.000 |

Dependent Variable: Days spent on ventilator.

## The PR model:

The estimates of the regression coefficients (B) and their statistical properties are shown in Table 11.6.

TABLE 11.6: Parameter estimates

| Parameter | B | Std. Error | 95% Wald CI | Sig. |
|---|---|---|---|---|
| (Intercept) | 0.546 | 0.196 | (0.162, 0.93) | 0.005 |
| [Diagnosis=1] | 0.843 | 0.411 | (0.037, 1.648) | 0.040 |
| [Diagnosis=2] | 0.205 | 0.162 | (-0.111, 0.522) | 0.204 |
| [Diagnosis=3] | 0[a] | | | |
| [Shock=0] | -0.348 | 0.175 | (-0.691, -0.004) | 0.047 |
| [Shock=1] | 0[a] | | | |
| DurHospStay | 0.069 | 0.007 | (0.056, 0.083) | 0.000 |
| (Scale) | 1[b] | | | |

Dependent Variable: Days spent on ventilator.
Model: (Intercept), Diagnosis, Shock, DurHospStay.
a. Set to zero because this parameter is redundant.
b. Fixed at the displayed value.

Using the results shown in Table 11.6, the model is written as

$$\ln(\mu) = 0.546 + 0.843 * (\text{Diagnosis=1}) + 0.205 * (\text{Diagnosis=2})$$
$$-0.348 * (\text{Shock=0}) + 0.069 * \text{DurHospStay}$$

On substituting the data on Diagnosis, Shock and DurHospStay for a new individual, we obtain the predicted value. This value is the expected duration of days on ventilator

## Predicted values:

With the options selected in the 'save' tab, we get the predicted and residual values for each record and a sample output is shown in Table 11.7.

TABLE 11.7: Predicted response and residuals of 'days on vent'

| S.No | Observed | Predicted | Residual |
|------|----------|-----------|----------|
| 1 | 5 | 4 | 1.038 |
| 2 | 5 | 3 | 1.564 |
| 3 | 0 | 3 | -2.792 |
| 4 | 3 | 4 | -1.236 |
| 5 | 5 | 6 | -0.588 |
| 6 | 4 | 3 | 0.550 |
| 7 | 2 | 2 | -0.431 |
| 8 | 5 | 5 | 0.135 |
| 9 | 0 | 3 | -3.219 |
| 10 | 18 | 10 | 8.277 |
| 11 | 12 | 4 | 8.038 |
| 12 | 4 | 3 | 1.395 |
| 13 | 0 | 3 | -2.992 |
| 14 | 4 | 3 | 1.208 |
| 15 | 4 | 2 | 1.723 |

The mean and S.D of residuals can be calculated and in general the mean will be close to zero. Thus Poisson regression can be run with either R or SPSS effectively.

In the following section we refer to some applications of Poisson regression.

## 11.5    Applications of the Poisson Regression model

Stefany *et. al* (2009) gave a detailed explanation about the analysis of the count data through Poisson regression. In their study, the focus was on observing the model's fit in handling '0' counts in the data. The application of this model was in estimating the number of alcoholic drinks consumed in terms of rating values of 'sensation making' and 'gender.'

Cameron and Trivedi (1998) emphasized the broad applicability of the Poisson regression model in diverse fields such as health services, economics, vital statistics and community medicine to mention a few.

Dobson (2002) detailed the practical utility of the PR model in estimating the death rate from coronary heart disease for smokers and non-smokers. A good nnumber of real life examples are given for better understanding of the theory and application of this model. Some more applications can be seen in Fox (2008) and Siddiqui *et. al* (1999).

Some interesting applications of Poisson regression are aimed at estimating the following.

a) Number of credit cards a person has, given the income and socio-economic status.

b) Number of mobile phones one has, given the level of employment, income etc.

c) Number of visits (per year) to a hospital by a person, given the health indicators and risk factor (this is actually a 'rate' and not count).

d) Number of women who gave birth to children even after completion of child bearing age.

e) Duration of hospital stay in a month.

f) Number of recreational trips to a lake in a year by a 'class of people.'

We end this chapter with the note that while logistic regression predicts a binary outcome, the Poisson regression predicts the counts of rare events.

---

## Summary

Poisson regression is special type of statistical regression model used to predict the outcome having a count data of rare events. It is a member of the class of log linear models and the model building is an iterative process. We have focused on the use of this model in the R and SPSS environments.

---

## Do it yourself (Exercises)

**11.1** Consider the data given in Illustration 11.1 and answer the following using SPSS.

a) Obtain the Poisson regression model.

b) Write a short note on the findings observed in parameter estimates from the output.

**11.2** Consider the data given in Illustration 11.3 and perform the following using R.

    a) Obtain the relative risk for each variable and interpret it.

    b) Obtain the predicted values using the Poisson model and test for model fit.

**11.3** Reconsider the data of Exercise 11.2 and fit a Poisson regression model using R with variable selection using the q-method with q=2 and also using the vector method.

**11.4** Perform the following in R.

    a) Generate a random sample of size 25 using rpois( ) function by taking values of $\beta_0$ and $\beta_1$ as 1 and 0.358 respectively.

    b) For the data generated, fit a Poisson model and comment on the goodness of fit.

**11.5** Generate random samples of size 50 for different values of $\mu$ to show that Poisson distribution tends to normal $\mu$ when large.

---

# Suggested Reading

1. Cameron, A.C and Trivedi, P.K. 1998. *Regression Analysis of Count Data*. $2^{nd}$ ed. Cambridge University Press.

2. Dobson, A.J. 2002. *An Introduction to Generalized Linear Models*. $2^{nd}$ ed. Chapman & Hall/CRC.

3. Fox, J. 2008. *Applied Regression Analysis and Generalized Linear Models*. $2^{nd}$ ed. Thousand Oaks, Ca: Sage.

4. Stefany Coxe, Stephene G. West and Leone S. Aiken (2009), The Analysis of Count Data: A Gentle Introduction to Poisson Regression and its Alternatives, *Journal of Personality Assessment*, 91(2), 121 –136.

5. Siddiqui O, Mott J, Anderson T and Flay B (1999), The application of Poisson random-effects regression models to the analysis of adolescents' current level of smoking. *Preventive Medicine*, 29, 92-101.

6. Greenwood, M., and Yule, G.U. 1920. An inquiry into the nature of frequency distributions of multiple happenings, with particular reference to the occurrence of multiple attacks of disease or repeated accidents, *Journal of Royal Statistical Society A*, 83, 255 –279.

# Chapter 12

## Cluster Analysis and Its Applications

| 12.1 | Data Complexity and the Need for Clustering | 219 |
|------|--------------------------------------------|-----|
| 12.2 | Measures of Similarity and General Approach to Clustering | 221 |
| 12.3 | Hierarchical Clustering and Dendrograms | 224 |
| 12.4 | The Impact of Clustering Methods on the Results | 229 |
| 12.5 | K-Means Clustering | 232 |
|      | Summary | 235 |
|      | Do it yourself (Exercises) | 235 |
|      | Suggested Reading | 237 |

> To understand God's thoughts we must study statistics, for those are the measure of His purpose.
>
> Florence Nightingale (1820 – 1910)

## 12.1  Data Complexity and the Need for Clustering

Cluster Analysis (CA) is a multivariate statistical tool procedure used in a context where a large number of sample subjects, each described by several variables (characteristics/attributes) have to be 'grouped' into two or more smaller sets called *clusters*.

The subjects for classification can be entities like banks, hospitals, districts/states or even individuals. We also refer to subjects as objects in some parts of the discussion.

When the data is observed on several subjects it is possible that some of the subjects exhibit *similarity* (in some sense, like closeness) not visible by simple observation. If identified, we can consider similar subjects as a *cluster* and understand the characteristics of members in the cluster. Such clusters are like *hidden colonies* of subjects which can be extracted by statistical methods. The aim is to discover clusters from complex data such that all subjects within a cluster are *similar* but each cluster is distinct (non-overlapping) from the other. These clusters often represent different populations which were mixed up in the data before clustering.

Clustering is different from 'classification' where the classes or groups (similar to clusters) are pre-defined and the job is to assign every individual to one of the two or more classes basing on the data observed about the individuals. The percentage of correct classifications is a measure of performance of the classifier.

Now in clustering we do not have pre-defined groups and we need to discover the clusters, if any. If the sample data has m-individuals, one trivial cluster is a single group of all individuals (possibly they are similar since they are relevant to the aim of the study!) or m-clusters where possibly all are different!

Modern decision making is *evidence based* and not based on beliefs or faith of individuals. All evidence is quantifiable and makes *data* whose size is *big* in some sense. Not only the number of records but also the number of dimensions (attributes), the data type and the velocity at which data is added to the storage, manifests in *big data* which is the focus of *data analytics.* In addition, there will be complicated relationships/linkages among the records as well as variables which leads to *data complexity.*

CA has many interesting applications such as grouping of psychiatric patients based on their symptoms, segmentation of genes having similar biological functions, grouping of localities like districts/states that are similar on the grounds of health care services, education, socio economic status etc. Identification of clusters helps in the effective management of services to the target groups.

CA is also an exploratory study and iterative computations are required before solving the problem. As such it is like an intermediate analysis rather than the end outcome (like p-value in tests of hypothesis). It is more a computer-intensive procedure that works with powerful algorithms in addition to providing *visualization* of cluster information to the end user. Liao *et.al* (2016) reported the application of CA to discover the patterns in the 'health care claims' of 18,380 CKD patients using retrospective data. CA is also known as *unsupervised learning* by researchers in the field of data mining and machine learning, as compared to *supervised learning* meant for classification into known groups.

There is a vast literature on the theory and application of CA. Johnson and Wichern (2009) discussed the background mathematical treatment of CA while Rencher (2002) provided sufficient discussion on CA with applications.

In the following section we examine some basic concepts of CA and learn how to process data to perform CA.

## 12.2 Measures of Similarity and General Approach to Clustering

The data for CA needs variables shown as columns and data records as rows (as done in a spreadsheet). For instance every row represents a subject (patient) for whom several characteristics are measured which are common to many subjects. Consider the data in Table 12.1 on the state-wise number of First Referral Units (Hospitals) in various States/UTs in India as on the $31^{st}$ of March 2017.

TABLE 12.1: State-wise details on First Referral Units (FRU) in selected states of India

| Identification Number | State/ UT | % of hospitals having | | | |
|---|---|---|---|---|---|
| | | More than 30 beds (X1) | Operation Theatre (X2) | Labour Room (X3) | Blood Storage/Linkage (X4) |
| 1 | Andhra Pradesh | 34.0 | 100.0 | 100.0 | 50.6 |
| 2 | Arunachal Pradesh | 71.4 | 92.9 | 92.9 | 92.9 |
| 3 | Assam | 92.0 | 92.0 | 92.0 | 92.0 |
| 4 | Bihar | 100.0 | 100.0 | 100.0 | 43.3 |
| 5 | Chhattisgarh | 70.7 | 78.7 | 100.0 | 82.7 |
| 6 | Goa | 25.0 | 6.3 | 6.3 | 6.3 |
| 7 | Gujarat | 100.0 | 88.3 | 100.0 | 100.0 |
| 8 | Haryana | 95.3 | 95.3 | 100.0 | 86.0 |
| 9 | Himachal Pradesh | 100.0 | 100.0 | 100.0 | 100.0 |
| 10 | Jammu & Kashmir | 74.7 | 94.9 | 100.0 | 52.5 |
| 11 | Jharkhand | 38.4 | 100.0 | 100.0 | 58.9 |
| 12 | Karnataka | 100.0 | 99.1 | 99.1 | 59.1 |
| 13 | Kerala | 100.0 | 100.0 | 100.0 | 100.0 |
| 14 | Madhya Pradesh | 100.0 | 100.0 | 100.0 | 95.0 |
| 15 | Maharashtra | 71.0 | 100.0 | 100.0 | 78.5 |
| 16 | Manipur | 85.7 | 85.7 | 85.7 | 14.3 |
| 17 | Meghalaya | 100.0 | 87.5 | 87.5 | 75.0 |
| 18 | Mizoram | 100.0 | 100.0 | 100.0 | 88.9 |
| 19 | Nagaland | 87.5 | 87.5 | 100.0 | 50.0 |
| 20 | Odisha | 55.4 | 100.0 | 100.0 | 100.0 |
| 21 | Punjab | 38.2 | 69.7 | 73.2 | 26.0 |
| 22 | Rajasthan | 100.0 | 94.8 | 100.0 | 62.1 |
| 23 | Sikkim | 100.0 | 100.0 | 100.0 | 100.0 |
| 24 | Tamil Nadu | 100.0 | 100.0 | 100.0 | 100.0 |
| 25 | Telangana | 100.0 | 100.0 | 100.0 | 56.6 |
| 26 | Tripura | 100.0 | 50.0 | 100.0 | 100.0 |
| 27 | Uttarakhand | 66.1 | 95.2 | 91.9 | 38.7 |
| 28 | Uttar Pradesh | 80.0 | 100.0 | 100.0 | 66.1 |
| 29 | West Bengal | 79.0 | 91.1 | 100.0 | 56.5 |
| 30 | A & N Island | 100.0 | 100.0 | 100.0 | 100.0 |
| 31 | Chandigarh | 100.0 | 75.0 | 100.0 | 100.0 |
| 32 | Dadra & Nagar Haveli | 100.0 | 100.0 | 100.0 | 100.0 |
| 33 | Daman & Diu | 100.0 | 100.0 | 100.0 | 100.0 |
| 34 | Delhi | NA | NA | NA | NA |
| 35 | Lakshadweep | 100.0 | 100.0 | 100.0 | 100.0 |
| 36 | Puducherry | 100.0 | 100.0 | 100.0 | 100.0 |

For the purpose of illustration, the State/UTs numbers 16 to 20 alone will be considered from Table 12.1.

Each variable represents one dimension, indicating the facilities available at the FRUs in the State/UT. We wish to identify the states that are similar with respect to these 4 dimensions and segment them into a few groups so that suitable policies can be framed basing on similarity. CA helps in achieving this objective. Since data was not available for Delhi, that record will not be considered in the analysis.

The goal of CA is to evaluate a *similarity metric* for all possible pairs of states and segment them into clusters so that a) all the states within a cluster are similar and b) any pair of clusters will be as distant as possible. One way is to define 35 clusters, one for each state and another way is to put all in a single cluster. Both decisions are extreme in a practical sense. So we need a rational way of identifying similar objects using the data on all the variables.

One commonly used measure of similarity is the *Euclidean distance* between objects A and B defined as

$$d_{AB} = \sqrt{(x_A - x_B)^2 + (y_A - y_B)^2} \tag{12.1}$$

where x and y refer to the data on the objects A and B respectively.

When each object has k variables $(X_1, X_2, \ldots, X_k)$, the expression in (Equation 12.1) takes the following form

$$d_{AB} = \sqrt{((x_{1A} - x_{1B})^2 + (x_{2A} - x_{2B})^2 + \ldots + (x_{kA} - x_{kB})^2)} \tag{12.2}$$

For instance, the distance between Manipur and Mizoram is

$$\sqrt{((85.70 - 100.00)^2 + (85.70 - 100.00)^2 + (85.70 - 100.00)^2 + (85.70 - 88.90)^2)}$$

$$= \sqrt{204.49 + 204.49 + 204.49 + 5565.16}$$

$$= 78.60$$

Similarly between Manipur and Nagaland it is only 38.54. So Manipur and Nagaland are more *similar* (lesser distance) than Manipur and Mizoram. Using the MS-Excel Array formula (Example: SQRT((I13:K13-I14:K14)^2)) we can find the distances easily.

Though the quantity d given in (12.2) is called *distance*, it has no units of measurement like meters for physical distance. It is a unit-free number and used as an index of proximity when data is expressed in different units of measurement. If we consider a subset of 4 states numbered 16 to 19, the distance matrix, also called *proximity matrix*, can be calculated as given in Table 12.2 (software like SPSS or R will produce this table).

When all the states are considered (instead of 5) we get a matrix of order (35 x 35) with all diagonal elements 0. It is also easy to see that similar to the correlation coefficient, the Euclidean distance is also symmetric.

TABLE 12.2: Proximity matrix

| State/UT | Euclidean Distance | | | | |
|---|---|---|---|---|---|
| | Manipur | Meghalaya | Mizoram | Nagaland | Odisha |
| Manipur | 0 | 62.41 | 78.60 | 38.54 | 93.12 |
| Meghalaya | 62.41 | 0 | 22.49 | 30.62 | 54.09 |
| Mizoram | 78.60 | 22.49 | 0 | 42.73 | 45.96 |
| Nagaland | 38.54 | 30.62 | 42.73 | 0 | 60.71 |
| Odisha | 93.12 | 54.09 | 45.96 | 60.71 | 0 |

In this example all four variables have the same unit of measurement viz., percentage and hence are comparable. When the measurement scales are different the rankings (lowest or highest) of the distances get altered.

One method of overcoming this issue is to standardize the data on each variable X by transforming into $Z = (x - m)/s$, where x = data value, m = mean and s = standard deviation of X. Rencher (2002) remarks that, "CA with Z score may not always lead to a *good separation* of clusters. However, Z score is commonly recommended."

Some algorithms like the one in SPSS use other scaling methods for standardization, e.g.,

- Transforming to a range of (-1 , 1)

- Transforming to a range of (0,1)

- Ceiling to a maximum magnitude of 1

- Transforming so that mean is 1

- Transforming so that standard deviation is 1

In addition to the Euclidean distance, there are other measures of similarity used in cluster analysis like a) squared Euclidian distance, b) Pearson's correlation, and c) Minkowski distance. Each measure of distance or transformation formula has its own merits and contextual suitability. Comparing them and assessing their strengths is beyond the scope of this book. However, for practitioners, one rule of thumb is to pick up two or more methods and choose the better outcome.

CA is applied for either *clustering of objects* (like individuals/districts/ hospitals) or *clustering of variables* (like demographic variables, biochemistry variables, income). Both objectives lead to smaller subsets of objects/variables which have similarity. The second one is much similar to factor analysis.

Unless stated otherwise, we concentrate on clustering of objects instead of variables.

There are two approaches for clustering, namely, a) hierarchical clustering and b) non-hierarchical clustering which are outlined in the following section.

---

## 12.3    Hierarchical Clustering and Dendrograms

Suppose there are $n$-objects and on each object $p$-variables are measured (we may also use categorical variables). Assume that we wish to detect k-clusters out of the $n$ objects. Then the number of possible clusters is approximately $[k^n/k!]$ which is $8.34*10^{15}$ for small values like k = 3 and n = 35 and it is practically impossible to list the clusters before selecting the best few. Hence several algorithms are available to search for the best clusters.

Hierarchical clustering is an iterative process in which clusters are formed sequentially basing on the distance between the objects. This is done in two ways as given below.

a) **Agglomerative Clustering:** In this method each object is initially kept in a separate cluster so that there will be as many clusters as objects. Based on the chosen similarity measure, an object is *merged* with another to form a cluster (agglomeration or putting together). This procedure is repeated until all the objects are clustered. This is a *bottom to top* approach.

b) **Divisive Clustering:** This is a *top to bottom* approach where all objects are kept in a single cluster which is sequentially split (segmented) into smaller clusters basing on the distance.

For simplicity and in keeping with common practice, we recommend the use of agglomerative clustering only.

The agglomerative hierarchical algorithm starts with a pair of objects having the smallest distance, which is marked as one cluster. The pair with the next smallest distance will form either a new cluster or be merged into the nearest cluster already formed.

There are different ways of forming clusters by defining proximity between objects, as outlined below.

a) **Single Linkage Method:** This is also known as *nearest neighbourhood* *method* in which clusters are separated in such a way that the distance between any two clusters is the *shortest distance* between any pair of objects within clusters.

b) **Complete Linkage Method:** This is similar to the single linkage method but at each step a new cluster is identified basing on the maximum or largest distance between all possible members of one cluster and another. Most distant objects are kept in separate clusters and hence this method is also called *farthest distance neighbourhood* method.

c) **Average Linkage Method:** In this method the distance between two clusters is taken as the *average distance* of all possible objects belonging to each cluster.

d) **Centroid Method:** In this method the geometric center (called a *centroid*) is computed for each cluster. The most used centroid is the simple average of clustering variables in a given cluster. The distance between any two clusters is simply the distance between the two centroids.

A dendrogram is a graphical representation of clusters basing on the similarity measure. It plays a vital role in visualization of cluster formation.

Here is an illustration of how the agglomeration method works.

**Illustration 12.1** Reconsider the data given in Table 12.1 on Rural Health Facilities. To keep the discussion simple let us select only 5 states numbered 16 through 20 for clustering. Clustering will be based on the distance matrix given in Table 12.2.

**Analysis:**

The steps of hierarchical clustering are as follows.

**Step-1:** Locate the smallest distance in the matrix. It is 22.49 (shown in boldface) between 17 and 18. So the first cluster will be C1 = {17, 18}.

| State/UT | 16 | 17 | 18 | 19 | 20 | Cluster | Min Distance |
|----------|------|------|------|------|------|---------|--------------|
| 16 | 0 | 62.41 | 78.6 | 38.54 | 93.12 | | |
| 17 | 62.41 | 0 | **22.49** | 30.62 | 54.09 | C1 = | |
| 18 | 78.6 | 22.49 | 0 | 42.73 | 45.96 | {17, 18} | 22.49 |
| 19 | 38.54 | 30.62 | 42.73 | 0 | 60.71 | | |
| 20 | 93.12 | 54.09 | 45.96 | 60.71 | 0 | | |

Record the new cluster and the distance as shown above. Remove the data in column 18 and rename the row 17 as C1. Proceed to step-2 with the reduced matrix.

**Step-2:** Locate the smallest distance from the reduced matrix. It is 30.62 between C1 and 19 and hence the next cluster is
C2 = {C1, 19}.

| | 16 | 17 | 18 | 19 | 20 | | |
|---|---|---|---|---|---|---|---|
| 16 | 0 | 62.41 | | 38.54 | 93.12 | | |
| C1 | 62.41 | 0 | | **30.62** | 54.09 | C2={C1,19} | 30.62 |
| 19 | 38.54 | 30.62 | | 0 | 60.71 | | |
| 20 | 93.12 | 54.09 | | 60.71 | 0 | | |

Record the new cluster and the distance. Remove the data in column 19 and rename the row C1 as C2. Proceed to step-3 with the reduced matrix.

**Step-3:** The next smallest distance is 54.09 between C2 and 20 and hence the new cluster is C3 = {C2,20}.

| | 16 | 17 | 18 | 19 | 20 | | |
|---|---|---|---|---|---|---|---|
| 16 | 0 | 62.41 | | | 93.12 | | |
| C2 | 62.41 | 0 | | | **54.09** | C3= {C2,20} | 54.09 |
| 20 | 93.12 | 54.09 | | | 0 | | |

Remove the data in column 20 and rename the row C2 as C3. Remove row 20 also because it is already clustered. Proceed to step-4 with the reduced matrix.

**Step-4:** The next smallest distance is 54.09 between C2 and 20 and hence the new cluster is C3= {C2,20}. The only states to be clustered are 16 and 17 but 17 is already contained in C3. Hence the final cluster will be C4 = {C3,16}.

| | 16 | 17 | 18 | 19 | 20 | | |
|---|---|---|---|---|---|---|---|
| 16 | 0 | **62.41** | | | | C4= {C3,16} | 62.41 |
| C3 | 62.41 | 0 | | | | | |

Now the clustering is complete with 4 clusters as listed below.

| Cluster | State Codes | Distance | States |
|---|---|---|---|
| C1 | {17, 18} | 22.49 | *{Meghalaya, Mizoram}* |
| C2 | {17,18,19} | 30.62 | *{Meghalaya, Mizoram, Nagaland}* |
| C3 | {17,18,19,20} | 54.09 | *{Meghalaya, Mizoram, Nagaland, Manipur}* |
| C4 | {17,18,19,20,16} | 62.41 | *{Meghalaya, Mizoram, Nagaland, Manipur, Odisha}* |

If we use software like SPSS the Cluster membership will be shown as below.

| Case | Cluster Number |
|------|----------------|
| 16:Manipur | 1 |
| 17:Meghalaya | 2 |
| 18:Mizoram | 2 |
| 19:Nagaland | 3 |
| 20:Odisha | 4 |

The dendrogram will be as shown in Figure 12.1.

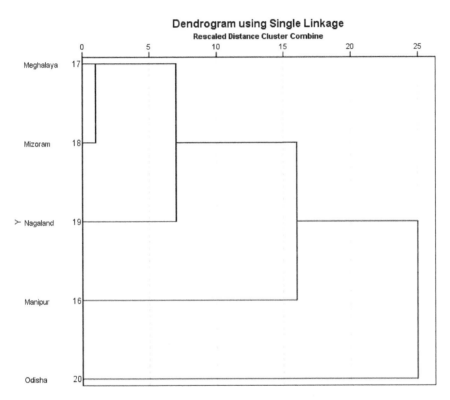

FIGURE 12.1: Dendrogram using SPSS.

By default SPSS produces the dendrogram with distance rescaled from 0 to 25. If we use R software we get the dendrogram in which the clusters are enclosed in rectangles marked in different colors for easy identification. For instance when opting for 4 clusters (k = 4) the dendrogram looks as shown in Figure 12.2.

**Cluster Dendrogram**

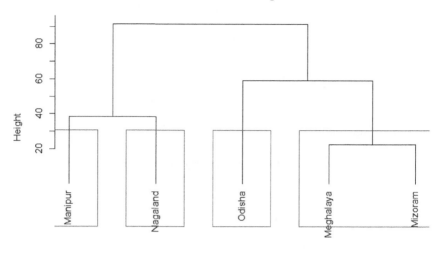

d
hclust (*, "ward.D")

FIGURE 12.2: Dendrogarm of Rural Health Facilities using R.

In a dendrogram each rectangular box represents a cluster and it is easy to see that this segmentation matches with the groups given in Figure 12.1.

As a post hoc analysis we can obtain the statistical summary (average) of the variables within each cluster as given in Table 12.3 to understand the performance of each clustering variable.

TABLE 12.3: Cluster-wise mean % for variables

| Cluster | States/ UT | X1 | X2 | X3 | X3 |
|---------|------------|-----|-----|-----|-----|
| 1 | Manipur | 85.70 | 85.70 | 85.70 | 14.30 |
| 2 | {Meghalaya, Mizoram} | 100.00 | 93.75 | 93.75 | 81.95 |
| 3 | Nagaland | 87.50 | 87.50 | 100.00 | 50.00 |
| 4 | Odisha | 55.40 | 100.00 | 100.00 | 100.00 |

**Remark-1**

The number of groups (k) into which the tree (dendrogram) has to be split is usually done by trial and error. Alternatively, we may fix the height (distance or any other similarity measure) as some value and try grouping. Using the *cutree* command in R we can also specify the number of clusters required.

In the following section we discuss the importance of the method of clustering on the formation of clusters.

---

## 12.4   The Impact of Clustering Methods on the Results

The number of clusters and the final measure of similarity depend on the choice of the 'distance measure' as well as the 'linkage method.' A blind approach to one of the available options may not yield the best solution.

One way out is to run with the available options and select the better outcome. Running a code in R for hierarchical clustering is one common approach used by data scientists.

In the following section we illustrate the use of R-code to draw dendrograms by different methods and compare the results.

**Illustration 12.2** Reconsider the data used in Illustration 12.1 with all 35 records (barring Delhi). Instead of single linkage suppose we use complete linkage or average linkage as the method in agglomeration.

**Analysis:**

The following steps help in drawing the dendrogram with clusters marked on the tree.

1. Open R-studio

2. Import data from MS-Excel file

3. Name it as Data1

4. Type the following R-code and press 'enter' after each command.

   a. `y <- subset(Data1, select = c(X1,X2,X3,X4))` # only numerical variable are chosen for clustering.

   b. `ds <- dist(y,method = "euclidean")` # this defines the distance matrix with output named 'ds'

   c. `hc<- hclust(ds, method = "single")` # this produces hierarchical clustering with output as 'hc'

   d. `plot(hc, labels = Data1$'STATE/UT')` # this produces dendrogram with labels for state/UT

   e. `rect.hclust(hc,k = 10, border = "red")` # this breaks the dendrogram (tree) in to 10 groups marked in red.

f. Press 'OK' to run the code and view the output.

The method option in 'ds' can be changed to 'Squared Euclidean' or 'correlation' and this produces different clustering.

The method option in 'hc' can be changed from single to i) complete, ii)average, iii) centroid, iv) median, or v) Ward and this will change the agglomeration method. Both the dendrogram and the grouping pattern changes when this method is changed.

Keeping the Euclidean method as fixed, if we change the agglomeration we get the three shapes of a dendrogram each grouped into 10 clusters as shown in Figure 12.3, Figure 12.4 and Figure 12.5. The method is indicated below the horizontal axis.

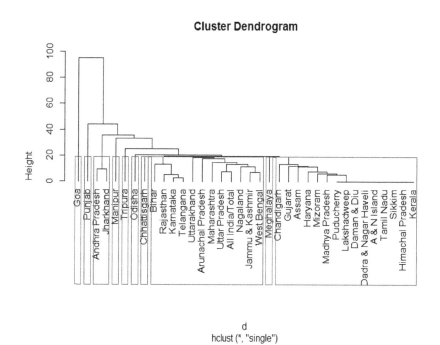

FIGURE 12.3: Hierarchical classification with single linkage.

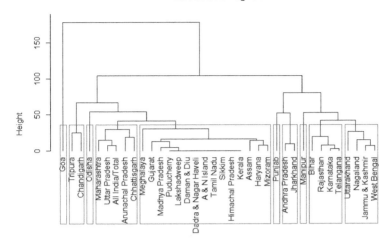

FIGURE 12.4: Hierarchical classification with complete linkage.

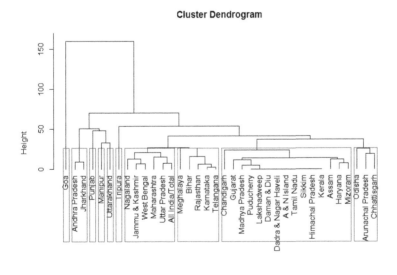

FIGURE 12.5: Hierarchical classification with average linkage.

The pattern of clustering changes with a) the variables selected for clustering, b) the number of clusters specified in the algorithm and c) the linkage method used. As such the clusters derived from the data or the dendrogram gives only a feasible segmentation of the objects. One has to try other patterns of clustering and choose the better one.

In the following section we use a non-hierarchical method by using centroids of data.

---

## 12.5   K-Means Clustering

This is a non-hierarchical clustering method and also known as *divisive Clustering*. The approach is different from the hierarchical method in the sense that the *centroid* of every object (item or data case) on all the variables is the criterion for clustering. The centroid is usually taken as the *mean vector* of all the p-variables. (K is used to denote the number of clusters and hence p used for the number of variables.)

In this method we have to specify the number of clusters (k) and the algorithm starts with a set of k-items from the data as a *seed*. At this stage each cluster contains only one item and the data values on the p-variables generate the centroid which is also called the *cluster center*. These centroids will be later updated when new items are included.

Given any two clusters the difference between the centroids is the criterion in the formation of clusters. The interesting feature in k-means clustering is that items which are included can also be removed from a cluster, when better objects are available for inclusion. This is not possible in hierarchical clustering.

Consider the following illustration.

**Illustration 12.3** Reconsider the data given in Illustration 12.1 with all 35 records. Instead of a hierarchical method we use the k-means method and cluster the hospitals.

If we use SPSS the following is the produce for performing k-means clustering.

a) Analyze→ Classify→k-means cluster.

b) The selections of variables (for clustering) and labels can be done with the options as shown in Figure 12.6.

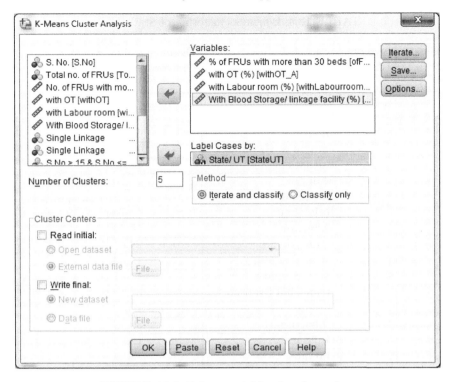

FIGURE 12.6: Select variables for clustering.

c) From the options window select ANOVA table and 'Cluster information' for each case.

d) Under the Save option select 'Cluster membership' and 'Distance from cluster center'.

Let us choose k = 5 and run the procedure. The output when rearranged produces the clusters as shown in Table 12.4. The distance of each item (state/UT) from the cluster center is an important information.

TABLE 12.4: k-means clustering of State/UTs

| Cluster | Size of Cluster | State/UT | Distance from Cluster Center | Cluster Centroid (Mean ± SE) |
|---|---|---|---|---|
| 1 | 2 | Tripura | 12.50 | 12.50 ± 0.0002 |
| | | Chandigarh | 12.50 | |
| 2 | 10 | Bihar | 15.40 | 17.91 ± 2.71 |
| | | Jammu & Kashmir | 13.07 | |
| | | Karnataka | 16.30 | |
| | | Manipur | 38.70 | |
| | | Nagaland | 7.69 | |
| | | Rajasthan | 17.75 | |
| | | Telangana | 15.43 | |
| | | Uttarakhand | 24.67 | |
| | | Uttar Pradesh | 18.63 | |
| | | West Bengal | 11.47 | |
| 3 | 19 | Arunachal Pradesh | 22.10 | 14.87 ± 2.02 |
| | | Assam | 8.310 0 0 | |
| | | Chhattisgarh | 30.43 | |
| | | Gujarat | 12.68 | |
| | | Haryana | 8.96 | |
| | | Himachal Pradesh | 10.22 | |
| | | Kerala | 10.22 | |
| | | Madhya Pradesh | 8.49 | |
| | | Maharashtra | 26.85 | |
| | | Meghalaya | 25.16 | |
| | | Mizoram | 10.01 | |
| | | Odisha | 37.64 | |
| | | Sikkim | 10.22 | |
| | | Tamil Nadu | 10.22 | |
| | | A & N Island | 10.22 | |
| | | Dadra & Nagar Haveli | 10.22 | |
| | | Daman & Diu | 10.22 | |
| | | Lakshadweep | 10.22 | |
| | | Puducherry | 10.22 | |
| 4 | 3 | Andhra Pradesh | 14.82 | 22.41 ± 5.50 |
| | | Jharkhand | 19.31 | |
| | | Punjab | 33.11 | |
| 5 | 1 | Goa | 0.00 | |

It can be seen from Table 12.4 that 19 states belonged to cluster-3 while 10 belonged to cluster-2. The cluster centroids are seen to be increasing from cluster-1 to cluster-5. A smaller value of the centroid indicates closeness/similarity and hence the first cluster has the smallest value. Cluster-5 has only Goa and hence it has zero distance from the centroid.

As a post hoc analysis, the k-means clustering produces information on the homogeneity of the mean vectors (of variables) across the clusters. More applications of cluster analysis using SPSS in the context of marketing research are discussed by Sarstedt and Moopi (2014).

We end this chapter with the observation that grouping of objects basing on several dimensions is a challenging job. Data scientists, particular, in business decision-making, use machine learning tools to perform and update solutions in real time.

## Summary

Cluster analysis is a vital tool in problems of pattern recognition. Grouping of objects basing on multidimensional data is a complex phenomenon but has potential applications in various areas like marketing research, psychology, facilities planning, drug discovery and gene expression analysis. The tools used in cluster analysis are mostly computer intensive and a great deal depends on data visualization. We have discussed two broad methods of clustering namely hierarchical clustering and k-means clustering with the help of SPSS and also R. Even though clustering is a popular method of machine learning it is often not the end outcome and the use of this tool needs trained statisticians in the study team.

## Do it yourself (Exercises)

**12.1** The following data refers to the distribution of states/Union Territories (UT) of India with respect to implementation of Hepatitis-B vaccine-CES 2009. The figures are percentage of children aged 12 –35 months who received the vaccine under the UIP Programme. (Source: https://nrhm-mis.nic.in/SitePages/Pub-FW-Statistics2015.aspx)

| S.No | States/Union Territories | Hep B at Birth | Hep_B1 | Hep_B2 | Hep_B3 |
|------|--------------------------|----------------|--------|--------|--------|
| 1 | Andhra Pradesh | 13.8 | 94.4 | 90.5 | 70.8 |
| 2 | Delhi | 63.1 | 72.8 | 68.5 | 64.5 |
| 3 | Goa | 45.2 | 76.7 | 81.4 | 81.3 |
| 4 | Himachal Pradesh | 47.1 | 76.5 | 73.5 | 71.0 |
| 5 | Jammu & Kashmir | 46.9 | 65.9 | 63.0 | 59.2 |
| 6 | Karnataka | 51.0 | 89.6 | 85.7 | 76.1 |
| 7 | Kerala | 68.4 | 85.9 | 83.2 | 81.5 |
| 8 | Madhya Pradesh | 12.3 | 30.3 | 27.8 | 24.3 |
| 9 | Maharashtra | 23.7 | 67.8 | 62.7 | 56.0 |
| 10 | Punjab | 32.9 | 87.5 | 85.2 | 82.3 |
| 11 | Sikkim | 1.4 | 90.9 | 81.9 | 79.3 |
| 12 | Tamil Nadu | 46.4 | 73.0 | 69.1 | 63.3 |
| 13 | West Bengal | 8.8 | 56.3 | 51.1 | 43.6 |
| 14 | UTs combined* | 47.3 | 75.1 | 72.1 | 65.5 |

Use a suitable clustering method and identify the states that are similar with respect to implementation of Hep B vaccine at each of the four levels (Hep B at Birth, Hep_B1, Hep_B2, He_B3) and compare the clusters.

**12.2** The following data refers to values of blood parameters in a hematology study on 75 patients who were primarily classified into 3 anemic groups. The variables are mean corpuscular volume (MCV, X1), Vitamin-B12 (B12, X2), Serum Homocysteine (SH, X3) and Transferrin Saturation (TS, X4).

| Anemic Group = 1 | | | | | Anemic Group = 2 | | | | | Anemic Group = 3 | | | |
|---|---|---|---|---|---|---|---|---|---|---|---|---|---|---|
| S.No | X1 | X2 | X3 | X4 | S.No | X1 | X2 | X3 | X4 | S.No | X1 | X2 | X3 | X4 |
| 1 | 80 | 370 | 10 | 13 | 26 | 67 | 155 | 13 | 4 | 51 | 75 | 299 | 19 | 58 |
| 2 | 61 | 331 | 38 | 4 | 27 | 73 | 129 | 14 | 5 | 52 | 127 | 96 | 50 | 52 |
| 3 | 91 | 1154 | 17 | 18 | 28 | 131 | 171 | 19 | 16 | 53 | 85 | 196 | 50 | 96 |
| 4 | 96 | 2000 | 21 | 4 | 29 | 69 | 241 | 21 | 3 | 54 | 87 | 148 | 20 | 39 |
| 5 | 64 | 567 | 13 | 5 | 30 | 98 | 290 | 8 | 4 | 55 | 100 | 283 | 21 | 27 |
| 6 | 82 | 329 | 8 | 19 | 31 | 65 | 213 | 21 | 4 | 56 | 95 | 254 | 19 | 95 |
| 7 | 79 | 630 | 9 | 8 | 32 | 61 | 297 | 14 | 6 | 57 | 126 | 142 | 50 | 97 |
| 8 | 66 | 1430 | 4 | 2 | 33 | 88 | 264 | 14 | 2 | 58 | 64 | 167 | 16 | 54 |
| 9 | 59 | 455 | 16 | 3 | 34 | 78 | 158 | 10 | 6 | 59 | 102 | 180 | 35 | 82 |
| 10 | 105 | 504 | 10 | 17 | 35 | 69 | 232 | 14 | 2 | 60 | 109 | 242 | 8 | 26 |
| 11 | 64 | 567 | 11 | 6 | 36 | 72 | 229 | 14 | 4 | 61 | 112 | 240 | 50 | 62 |
| 12 | 72 | 437 | 13 | 8 | 37 | 84 | 203 | 21 | 12 | 62 | 55 | 271 | 23 | 90 |
| 13 | 99 | 410 | 6 | 14 | 38 | 56 | 277 | 9 | 6 | 63 | 135 | 115 | 50 | 60 |
| 14 | 63 | 515 | 6 | 8 | 39 | 104 | 194 | 21 | 10 | 64 | 109 | 108 | 28 | 87 |
| 15 | 65 | 1053 | 5 | 3 | 40 | 98 | 197 | 23 | 7 | 65 | 77 | 169 | 24 | 81 |
| 16 | 72 | 801 | 7 | 4 | 41 | 80 | 135 | 25 | 11 | 66 | 136 | 86 | 32 | 102 |
| 17 | 76 | 853 | 8 | 10 | 42 | 80 | 243 | 24 | 3 | 67 | 76 | 250 | 16 | 56 |
| 18 | 95 | 374 | 4 | 12 | 43 | 105 | 175 | 27 | 16 | 68 | 142 | 171 | 17 | 44 |
| 19 | 115 | 643 | 24 | 19 | 44 | 52 | 277 | 9 | 4 | 69 | 126 | 160 | 19 | 27 |
| 20 | 67 | 375 | 10 | 1 | 45 | 58 | 213 | 19 | 5 | 70 | 100 | 283 | 9 | 21 |
| 21 | 64 | 809 | 9 | 4 | 46 | 126 | 105 | 33 | 3 | 71 | 103 | 150 | 18 | 39 |
| 22 | 82 | 345 | 24 | 6 | 47 | 64 | 257 | 10 | 3 | 72 | 123 | 83 | 22 | 207 |
| 23 | 57 | 363 | 10 | 5 | 48 | 64 | 279 | 10 | 3 | 73 | 113 | 70 | 46 | 21 |
| 24 | 87 | 397 | 15 | 11 | 49 | 63 | 250 | 16 | 5 | 74 | 93 | 187 | 9 | 20 |
| 25 | 88 | 851 | 15 | 7 | 50 | 89 | 250 | 50 | 2 | 75 | 109 | 145 | 24 | 27 |

Perform k-means clustering and identify the clusters. Compare them with the primary anemic groups and comment on the results.

**12.3** Consider the data given in Exercise 12.1. Apply different distance measures and comment on the clusters derived with hierarchical clustering.

**12.4** For the data given in Exercise 12.2, derive the cluster centroids and comment on the proximity of clusters.

---

# Suggested Reading

1. Johnson, R.A., & Wichern, D.W. 2014. *Applied multivariate statistical analysis*, 6th ed. Pearson New International Edition.

2. Alvin.Rencher(2002),Methods of Multivariate Analysis,Second Edition,Brigham Young University,John Wiley & Sons.

3. Marko Sarstedt and Erik Moopi 2014, A Concise Guide to Market Research-The Process, Data, and Methods Using IBM SPSS Statistics, Springer

4. Brian S. Everitt, MorvenLeese, Sbine Landau and Daniel Stahl. 2011, Cluster Analysis, John Wiley & Sons, Ltd

5. Liao et. al., 2016. Cluster analysis and its applications to health care claims data: A study of end-stage renal disease patients who initiated hemodialysis, *BMC Nephrology.* 17 (25), 1-46.

# Index

Additive Effect, 56
Adjusted F-ratios, 97
Adjusted R-Square, 56, 120
Adjusted Relative Risk, 198
Agglomerative, 224
ANCOVA, 61
ANOVA, 25, 51
attach(), 208
AUC, 144, 179
Average Linkage, 225

Baseline Value, 59
Beta Coefficients, 125, 131, 159
Between Subject Factor, 96, 110
Between-Subjects Effects, 59, 63
Big Data, 220
Binary Classification, 169
Binary Logistic Regression, 170
Biomarker Panels, 150
Biomarkers, 135, 144, 148, 149, 169
Bonferroni Adjustment, 98
Bonferroni Limits, 39
Box & Whisker Plot, 18, 100
Box's M-Test, 76, 78

Canonical Correlation Analysis, 158
Categorical Covariates, 177
Categorical Factors, 61, 83
Censored, 187, 189, 191, 194, 196
Centroid Method, 225
Classification Problem, 26, 136
Classification Table, 179
Classifiers, 136–138
Clinical trials, 186

Cluster Analysis, 26, 219, 223, 235
Cluster center, 232, 233
Coefficient of Determination, 11
Cofactors, 106
Collinearity, 125
Commensurate, 107
Complete Linkage, 225, 229, 231
Composite Classifier, 136, 148
Composite Score, 158, 161, 162
Conditional Probability, 169
Confidence Coefficient, 37, 40, 46
Confidence Interval, 8
Confidence Limits, 127
Correlation, 10, 11, 115, 131, 133
Count Data, 205
Count Data Regression, 206
COUNTIFS( ), 142
Covariance Matrix, 8–10, 27, 44
Covariates, 57, 61, 176
Cox Proportional Hazard, 195, 199
Cox Regression, 193
Cumulative Hazard Rate, 196
Cutoff Value, 137, 140, 142, 143, 147, 150
Cutree, 228

Data Analysis Pak, 15
Data Visualization, 16, 18
Dendrogram, 224, 228
Deviance, 207, 209, 211, 212
Discriminant Analysis, 136
Discriminant Score, 155
Distance Measure, 229
Distribution-free Methods, 65
Dummy Variable, 87

Duncan's Multiple Range Test, 52
Dunnett's Test, 52

Eigen Value, 76, 158
Equidispersion, 206
Euclidean Distance, 154, 164, 222, 223
Explanatory Variables, 118

Factor Analysis, 26
Factor Means, 59
False Negative, 138
False Positive, 138, 144, 151
Farthest Distance Neighbourhood, 225
Fisher's Linear Discriminant Function, 154
Force of Mortality, 193
Forward Conditional, 175, 176
Forward Selection, 211
Friedman's ANOVA, 68
FWDselect, 211

General Linear Model, 55, 82
Generalized Linear Models, 206, 213
Generalized Sample Variance, 10
glm( ), 208, 212
Gold Standard, 137
Goodness of Fit, 207
Greenhouse-Geisser, 93, 97
Group Membership, 173, 175, 178
Grouping Factor, 96

Hierarchical Clustering, 224, 225
Hotelling's T-Square, 32, 33, 74, 82
Hotelling's Trace, 76, 92, 97
Huynh-Feldt Method, 93
Hypothetical Mean Vector, 31

Incomplete data, 187
Inter Quartile Range, 18
Interactions, 55
Intercept, 57

K-Means Clustering, 232

Kaplan–Meier, 189, 190, 192, 199
Kolmogorov-Smirnov Test, 128, 157
Kruskal Wallis Test, 65

Least Squares, 116, 119
Levene's Test, 60, 92
Likelihood Ratio, 139
Linear Discriminant Analysis, 136, 153, 157, 164–166
Logistic Regression, 136, 169, 176
Long-Rank test, 193
Longitudinal Markers, 149, 150

Machine Learning, 136
Main Effects, 55
MANOVA, 25, 73
Marginal Influence, 61
Matrix Scatter Plot, 16, 27
Mauchly's W test, 92, 111
Mean Vector, 8, 10, 12, 23, 24, 31, 32
Method of Lower Bound, 93
Minkowski Distance, 223
Misclassification, 136
Model Fit, 121
Multicollinearity, 126
Multifactor ANOVA, 52
Multiple Comparison Test, 67
Multiple Correlation Coefficient, 119
Multiple Linear Regression, 25, 116, 154
Multivariate Analysis, 3
Multivariate Normal Distribution, 23

Nagelkarke R-Square, 177, 184
Nearest Neighbourhood, 225
Negative Predictive Value, 139
Non-hierarchical Clustering, 224, 232

Odds Ratio, 175, 178, 182, 196, 197
One-way ANOVA, 52
Outliers, 17

Paired sample Test, 44
Paired t-test, 43
Pairwise Comparison Test, 52
Pairwise Differences, 92
Parallel profile, 107, 110–112
Parameter Estimates, 86
Partial Correlation Coefficient, 121
Pillai's Trace, 76, 78, 79, 82, 84
Poisson Regression, 205–208, 213, 214, 216
Positive Predictive Value, 139
Post-hoc Analysis, 34, 67
Predicted Membership, 158, 160
Predictor Variables, 174, 206
Principal Components, 25, 26
Product-limit Estimate, 187
Profile Analysis, 106, 107, 109, 113
Profile Plot, 107, 112
Prognostic Factor, 107, 109
Proximity Matrix, 222

q-method, 211, 218
Qualitative data, 6, 7
Quantitative data, 5, 6

R-Square, 56, 118, 120, 123
Random Effects Model, 68
Random Error, 116
Random Factors, 57
Rao's Paradox, 32
Real Statistics, 15, 33, 76
Regression, 115, 116, 118–120, 122–124, 159
Regression Coefficient, 86, 116, 118, 119, 123–125, 131
Regressors, 118
Relative Importance, 125
Relative Risk, 193, 196, 198, 199, 209, 218
Repeated Measures, 91
Residual, 127, 208, 213, 215
Residual Deviance, 210
Residual Life, 185

RMANOVA, 25
ROC Curve, 143
Roy's Largest Root, 76
Rule of Classification, 172

Scatter Diagram, 19–21, 25
Scatter Matrix, 21
Scheffe's Test, 52
Sensitivity, 138, 139, 142
Separation, 223
Similarity Metric, 222
Simple Linear Regression, 116
Single Linkage, 224, 229
Specificity, 139, 142
Sphericity, 92, 97, 102, 109, 113
SPSS, 156–159
Standard Error, 7, 8, 119
Standardized Coefficients, 159
Stepwise Regression, 25, 120, 124, 126
Subset Regression, 211
Supervised Learning, 26, 136
Survival Analysis, 185, 186, 188, 193–195, 199
Survival Function, 187–189, 192, 200, 201
Survival Plot, 189
Survival Probability, 196
Survival Time, 186–190, 192, 200, 201

Tests Between Subjects, 82
Time-to-Event, 186
Tolerance, 125, 126
Trend Analysis, 98
True Negative, 138, 139
True Positive, 137, 141, 151
Two-way ANOVA, 68
Type-I Error Rate, 32

Unexplained Variation, 59
Univariate Analysis, 3
Unstandardized Coefficient, 124, 159
Unstandardized Residuals, 127
Unsupervised Learning, 26

Vector Method, 212, 218

Visualization of cluster, 225

Wald Chi-square, 215

Wilk's Lambda, 75

Within Subjects Factor, 96